THE CURIOSITY GENE

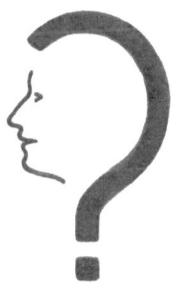

ALEXANDROS KOURT

Copyright © 2017 Alexandros S. Kourt

All rights reserved.

ISBN-10: 1547216956
ISBN-13: 978-1547216956
Library of Congress Control Number: 2017909259
CreateSpace Independent Publishing
Platform, North Charleston, SC

www.curiositygene.com

Artwork, cover & interior design: ASK.

THE CURIOSITY GENE

On the Origin of Humankind by Means of Intrinsic Motivation

CONTENTS

PREFACE   vii

PART ONE: CRACKING THE CODE OF HUMAN EVOLUTION   1

    1.1 A Large Brain Dilemma 2
    1.2 Competing Theories 10
    1.3 The Missing Link In Human Evolution 19

PART TWO: THE ORIGIN OF CURIOSITY   31

    2.1 An Elusive Drive: *Defining Curiosity* 32
    2.2 Of Chimps And Men 43
    2.3 War: *Curiosity's Violent Birth* 50

PART THREE: THE CURIOSITY GAME   79

    3.1 Pretraumatic Stress: *Remnants of Curiosity's Violent Beginnings* 80
    3.2 From War to Wonder: *The Demilitarization of Curiosity* 97
    3.3 Why Only Us? 112

PART FOUR: CURIOSITY RISING   125

    4.1 The Greatest Survival Tool On Earth 126
    4.2 The Unreasonable Minority: *Evolutionarily Stable Strategies* 139
    4.3 A Happiness Dilemma 147

NOTES   159

INDEX   171

ABOUT THE AUTHOR   177

PREFACE

Throughout the ages, people have asked perhaps the ultimate question: How did we get here? Among its connotations, there is the question of how the universe came to exist. There is also the question of how life began from the ubiquitous primordial soup. Then, consuming religion and science alike, there is the more personal question of how our species came to be. Even presupposing the existence of a universe with a diverse array of life, how on earth—excusing the pun—did such a unique and intelligent species as humankind come into existence?

With the advent of the theory of evolution, our species took a tremendous leap forward in answering this question, but it remains only a partial answer since we are so unique. We alone evolved such incredible intelligence...

Why only us?

We are the only species that gets to ask this question, of course. No surprise there, since only humans have sentences. We are the only creature smart enough to research and debate possible answers but what if our very desire to ask such questions is the reason for our smarts?

> *IS CURIOSITY ITSELF SOMEHOW RESPONSIBLE FOR OUR UNIQUE EVOLUTION INTO THE SMARTEST CREATURE ON THE PLANET? AND IF SO, HOW?*

In part one, *Cracking The Code of Human Evolution*, we will see how curiosity forms a missing link in the story of human evolution. We will see how we possess curiosity uniquely throughout the animal kingdom, at least in degree if not in form. Without it, we would still be ape-like and not have evolved into this present, intelligent form. We would be chimp, not human—a bold claim, to be sure. If indeed this is the case, it begs two further questions.

> FIRST, FROM WHENCE CAME THIS
> THING CALLED CURIOSITY?

In part two, *The Origin of Curiosity*, we will examine a momentous event of 2.3 million years ago as a likely "smoking gun" that set off a chain of circumstances in which our innate and insatiable form of curiosity appeared, stamped onto our DNA to make us truly unique in the animal kingdom. As with physical characteristics, desires and drives such as curiosity must likewise be products of evolution, which we will investigate for the first time.

> SECOND, IF CURIOSITY IS
> FUNDAMENTAL TO OUR EVOLUTION
> AND DEFINES OUR SPECIES, DOES IT
> NOT HOLD THE KEY TO HAPPINESS
> AND PERSONAL FULFILLMENT?

Towards the end of the book, we will illustrate the surprising road to happiness and fulfillment that highly curious individuals would be well-advised to pursue, or else risk feeling an enduring discontent tied to the events of 2.3

million years ago. For some, the curiosity gene can render contentment an elusive goal.

I believe that an understanding of our unique evolution has the ability to provide tremendous peace of mind and objectivity in an oft-confusing modern world. Emotions and feelings can be better understood given an awareness of our multimillion-year journey conditioned towards tribal life. With rates of depression and anxiety at record levels—350 million people suffering from depression globally according to the World Health Organization and one in four US teens suffering from generalized anxiety disorder according to the National Institute of Mental Health—something is possibly amiss.

For only a thin sliver of time, on geological timescales, has our emotional system been exposed to social media, processed foods, drugs, television, commercials, texting, internet surfing, politics, and career choices. Each modern element, while potentially adding comfort or entertainment to daily life, also preys on our attention and desire, resulting in potential conflicts. Seeing each feeling or motivation in its grander historical context can become a valuable tool to promote emotional intelligence of which the first step, as Daniel Goleman frequently wrote, is self-awareness and understanding. To that end, *The Curiosity Gene* touches upon emotions with an evolutionary backdrop to provide a deeper context on how this intrinsic drive can impact our personal happiness.[1]

My study of curiosity began quite unexpectedly while at the University of Manchester, England, in the throes of a computer science thesis. The specific topic under review involved logic circuits, the neurons of computers. What struck me about these computer building blocks was their similarity in function to certain aspects of the brain. This comparison, in turn, led me to this simple idea: If the brain is a computer, why not apply computer theory to understanding it and, in particular, its evolution? While this may sound reasonable, perhaps even obvious, it has not been a common practice in the field of anthropology. This book, among other things, aims to help change that.

Applying computer theory need not be complex or involve overwhelming jargon. Quite the contrary. During my twenty years building large-scale software systems in London and New York for some of the biggest companies in the world, a common theme emerged—namely that the simplest solution is often the best. While this is a well-known tenet called *Occam's razor*, named in honor of the fourteenth century English philosopher and friar William of Ockham (*Occamus* in Latin) and oft-quoted by Einstein, it is especially relevant in the software industry. Software developers have long-since come to understand that simpler solutions tend to be the best. They are relatively bug-free and efficient, so long as they are true to purpose, of course. A modern, technological mindset can and indeed should simplify our understanding of the brain while providing greater insight. With that, as we will soon see, light can be shed on brain evolution.

The brain is indeed an evolved organ. Stating the obvious, by better understanding something, we can better understand its evolution. If the brain is a computer, what hardware and software does it comprise? Examples of hardware in the technology industry include laptops, tablets, and smartphones. Their most fundamental component is the computer chip, also known as a CPU, silicon chip, or integrated circuit. Examples of software include Microsoft Word, smartphone apps, web browsers, and a special layer of core software called the OS (or "operating system"). The brain is not made of chips and wires but has similar elements. It has neurons, loosely analogous to hardware, and arrangements of electrical signals plus chemical signatures amid those neurons, loosely analogous to software.

Hardware and software interact continually, of course, in modern computer systems. This results in intricacies that become more apparent in large-scale systems, where multiple independent tasks are running simultaneously, each interacting with the others. Writing and enhancing large-scale computer systems is a rewarding but occasionally painful and frustrating endeavor.

Allowing computer scientists to make sense of it all is the concept of an *algorithm*. Google, founded in 1998 by Larry Page and Sergey Brin, has become one of the largest companies in the world with a market capitalization of over $600 billion, more than the GDP of most countries—all but the top twenty in fact! A tremendous reason for its success is not simply computer technology in general but specifically its use of algorithms. Similar to a strategy, an algorithm is broadly defined as a step-by-step procedure for solving a problem. A common example is a map app, which includes an algorithm to determine the best route between two addresses given data on available roads and speed limits. Such an algorithm can be challenging to create when the average speeds along the various stretches of road are impacted by congestion, stop signs, and traffic lights affecting optimal routes in intricate ways. Map algorithms offered by competing firms historically tended to be sub-optimal, allowing Google's continued rise to success. Other examples of algorithms or strategies, albeit far more humdrum, include how to make a cup of tea, pack a suitcase, or get dressed in the morning. Algorithms of various sorts are all around us. They are fundamental to the effectiveness of computer systems and everyday processes alike.

By focusing on algorithms instead of computer components, engineers have a more meaningful language in which to discuss software solutions. They provide a crucial level of abstraction. Likewise, focusing on mental algorithms instead of neurons and lobes can help decipher the brain's inner workings. Human behavior and desires can then be better understood, along with their origins.

Curiosity is itself an algorithm, a strategy, defined in part two. It is a mental process akin to software running on a computer or smart device. In a sense, a curiosity app is running inside our heads, even without our conscious knowledge. It motivates us, getting us to think and act in varying ways though not necessarily at the exclusion of other apps in our heads. Curiosity works alongside other mental processes, forming a mosaic of mental activity akin to multiple apps running on our smartphones.

Sometimes curiosity compliments other activities, driving us to improve existing processes, such as improving the way we work. At other times, curiosity conflicts with other processes, such as distracting us from perhaps less interesting work. Just as one app can drain CPU resources from other apps and frustrate us in what we are trying to accomplish, so too can conflicts occur between the curiosity app and other mental processes as they compete for our attention and neural resources. One of the great challenges in selecting a career, for example, is finding one that is both financially rewarding and mentally stimulating. Indeed, one of the great and unrelenting challenges in life is balancing competing demands and desires.

If curiosity is an app inside our heads, it is certainly one that profoundly impacts our daily lives. It is not only responsible for all the science and technology that we have come to enjoy, it also deeply affects many aspects of what we do, think, or even feel. Understanding the brain with that in mind—excusing yet another pun—can have profound implications on our understanding of how we got here.

THE CURIOSITY GENE

PART ONE

CRACKING THE CODE OF HUMAN EVOLUTION

1.1 A Large Brain Dilemma

It is fully one hundred and fifty years since Charles Darwin published his pivotal work, *On the Origin of Species by Means of Natural Selection*, introducing the world to the theory of evolution. Since then, there have been tremendous leaps in science and human understanding. We have documented over a million species on the planet. We have decoded the human genome. Our advancements have only strengthened our confidence in the theory of evolution, of course, but one gaping hole remains. How did we evolve?[2]

It may seem hard to believe that human evolution is considered a modern mystery, but a quick search on the internet will help clarify the situation. If you perform the following search…

Google | How did humans evolve? | 🔍

…you would expect to find a satisfactory explanation of how humans evolved. You would expect to read how and why our lineage diverged from the rest of the animal kingdom to result in our tremendous intelligence. Unfortunately, you would find no such explanation. Something is missing from our understanding of how we got here. Curiosity is that something. We will soon see how.

While evolution is behind our appearance along with other species, the scientific community is perplexed by one glaring problem. If intelligence provides a survival edge, why do all creatures not sprint forward to evolve it? Given millions of species over hundreds of millions of years, why only us? One hundred and fifty years of Darwinism has left us dumbfounded why we alone evolved such a large, powerful brain. What made our ancient ancestors different? Why did the various species of cat, dog, bear, elephant, pig, cow, horse, rat, rabbit, deer, and armadillo not also evolve a large brain despite each predating our own ape-like ancestors by millions of years?

Following your internet search, you would read that apes evolved into Australopithecus—a genus of species perhaps best described as upright chimps—and then various species of the genus Homo, culminating in us, Homo sapiens. However, you would not read how or why.

There is no shortage of research on our evolution, of course. The problem involves explaining how we are different. To this point, any explanation of human evolution could equally apply to other creatures. For instance, the classic explanation of how our large brain evolved is that intelligence provides a survival edge. As the theory goes, natural selection works its magic to ensure each generation, on average, inherits a larger, smarter brain due to this survival edge, but the same reasoning should apply to all species, not only our ancestors. What is missing?

One would imagine that an explanation ought to be glaring given how glaring is our relative intelligence across the animal kingdom. Our uniqueness is not exactly subtle. We are the dominant species on the planet, occasionally laying other species in our wake. We live in houses, sheltered from the elements. We make use of written communication that can pass knowledge down through generations. We have cars, airplanes, iPads, and go to the toilet in private, usually at least. We have liquid soap!

Given the solid foundation laid down by Darwin followed by over a century and a half of scientific research and technological advancement, you might feel rightfully disenfranchised upon failing to read a satisfactory answer to your internet search question. While intelligence is clearly a defining attribute of our species, why only we evolved it in spades remains a mystery. Other creatures certainly had sufficient time, the dinosaurs having been around for over 150 million years, almost forty times longer than us and our Australopithecus ancestors *combined*.

This omission is not through lack of effort by the scientific community, but the problem is, Darwin's theory is just too good. Whatever explanation has been posited for our brain's evolution could equally apply to other species, yet we alone became divergent. We alone experienced accelerated brain growth while other creatures did not.

Relative to our stature, our brain is far bigger than any other creature alive or that ever lived. It seems that large brains relative to body mass, or a high *encephalization quotient* to use the official terminology, is not the norm in nature. During the latter half of the twentieth century, scientists set about trying to explain why so few creatures experienced accelerated brain evolution…and they succeeded. Scientists discovered an overarching fact about brain evolution:

> *EVOLVING A LARGE BRAIN IS INCREDIBLY DIFFICULT.*

While intelligence has given us an obvious survival edge, large brains come at a high cost. Indeed, a large brain has the potential to be a burden, negatively impacting our chances of survival in many ways. There exists a veritable laundry list of challenges.

First, gray matter is incredibly energy- and nutrient-hungry. Pound-for-pound, the brain is the most energy-hungry organ in the body, eclipsing even skeletal muscle. Twenty percent of our calorific intake goes towards feeding the brain despite it being only two percent of our body mass. For primal forest dwellers, competing to obtain sufficient quantities of food was job number one. Natural selection over extended periods of time is a game of probabilities. Magnifying dietary demands tips the scale aggressively in the wrong direction. Other things being equal, creatures with smaller, more compact brains would have greater chances of survival by burning significantly less fuel in the form of food.[3]

Second, the brain uses tremendous amounts of oxygen, adversely impacting our stamina. Creatures with larger brains would be at greater risk from prey or rivals. When hunting their own prey for a meal, the reduction in stamina would again tip the scales in the wrong direction, in favor of the prospective prey and increasing the likelihood of a lost meal.

Continuing down the laundry list, the cellular mass of the brain puts extra strain on the immune system, leaving a creature at greater risk from bacteria and disease. Having a large brain also makes it more difficult to regulate body temperature, leaving a creature at greater risk during extreme shifts in ambient temperature. The sheer weight of the head means that a larger-brained creature would tend to be slower and less maneuverable. As with reduced stamina, the added weight creates challenges when chasing prey or, in turn, avoiding becoming another creature's prey, which we generally prefer to do! A larger brain is also far more vulnerable. The contraption on our shoulders must be kept safe from trauma during the eons before helmets and CAT scans came into being. Simply falling off a tree, slipping from a wet rock, or tripping over a root would suddenly add critical percentage points to the likelihood of catastrophic failure in the evolutionary relay race.

Finally and perhaps most ominously, a large head means that childbirth is more dangerous, both for child and mother. While we, in the developed world, have become accustomed to modern medicine to keep maternal and infant mortality rates in check, this is a relatively recent development if we consider multimillion-year timelines. Even today, in developing countries such as Afghanistan, Somalia, and Nigeria, it is estimated that one birth in every fifty will result in the death of the mother. Likewise, infant mortality rates top out at a whopping ten percent according to the Central Intelligence Agency's *World Factbook*. These mortality rates represent significant natural selection overheads right out of the gate, during the reproductive imperative of birth itself.[4]

Small brains imply fewer birth complications and longer gestation periods resulting in more robust newborns. Modern humans have evolved shorter gestational periods out of necessity to liberate the baby of its motherly cocoon before the head gets too large for vaginal birth to be a possibility. The cocoon would eventually become a coffin, to paint a grotesque picture. As a result, newborn humans are extremely vulnerable and dependent individuals compared with other newborn mammals. We have all seen images of newborn deer or horses getting to their feet and walking within minutes of birth. For humans, both mother and child unknowingly suffer from a large brain dilemma.

Our brain is our defining attribute. When we speak of explaining human evolution, we strive to explain the brain's evolution to its current 85 billion neuron composition. Other characteristics of our species are either shared with other mammals or circumstantial to our intellect. Our speech evolved due to the brain's capacity for speech. Many other creatures have audio communication but with speech centers typically less complex than ours. Other creatures have social structures and group dynamics though our brain power simply allows us to take things to another level. Other creatures, it is safe to assume, experience love. Many species of bird mate faithfully for life, such as the bald eagle and California condor, and countless species go through extended periods of mourning, presumably heartbroken, should a partner die. We evolved hand dexterity, complete with opposable thumbs, due to our mental capacity to take advantage of this faculty, though other creatures have skeletal digits. Neither speech, socialization, love, nor hand dexterity define us uniquely other than circumstantially to brain power.

The size of the brain, it would appear, is our defining attribute among these characteristics. What, then, of rumors that only ten percent of our brain is utilized? Quite simply, these are false rumors of an urban legend made believable through repetition and Hollywood movies like *Lucy*. Head trauma research has revealed that no area of the brain can be destroyed without impairing a mental capability, let alone ninety percent of the brain being damaged without consequences. Furthermore, the latest neuroscience, through experimentation, has illustrated that all areas of the brain react following electrical stimulation.

In seeking an origin to the notion of an underutilized human brain, it quickly becomes clear that the self-help industry going back over one hundred years has touted the notion of underutilized potential in order to encourage customers to purchase self-help books and seminar tickets. At some point, the concept of underutilized potential was misinterpreted as a clinically underutilized brain. This false notion was solidified when one of the highest-selling self-help books of all time, Dale Carnegie's *How to Win Friends and Influence People*, asserted that only ten percent of the brain was used in the preface authored by William James. While thoroughly false and misleading, the assertion was extremely authoritative.[5]

Logically speaking, the evolution of any unused gray matter, let alone nine-tenths of the brain's mass, would go against everything we know about natural selection. As mentioned, pound-for-pound, the brain is the most energy-hungry organ in the body, requiring over twenty percent of the calorific energy we consume through food. A large brain is not always a good thing. Other creatures did not evolve a large brain, despite the obvious advantages of intelligence, since a large brain can be a killer drawback. A large brain is expensive and can inhibit survival. Any unused excess would necessarily be weeded out by natural selection. Only by denying one hundred and fifty years of Darwinism can we defend the theory that most of our brain is unused matter in the form of neurons simply sitting idle as untapped potential. It should be considered a monumental feat of bravery to quote such a statistic outside the confines of a science fiction movie given the rotational impact to Darwin's remains. He would simply turn in his grave.

The many drawbacks of a large brain equate to profound evolutionary forces for *less* brainpower. In essence, these are driving forces toward a *smaller* brain despite the obvious advantages of intelligence. It then becomes a question of how to outweigh those drawbacks if a large brain is ever to evolve and tip the scales in the opposite direction. Given how few of the millions of species on the planet historically ever evolved a large brain, the scales are clearly tipped firmly on one side, usually. Evolution typically favors a sleek, streamlined, lightweight brain. Countless species did not overcome those drawbacks but our ancestors somehow did since clearly we exist! We indeed evolved, presumably through similar biological processes as all other creatures. If evolution arrived at this thing called humankind, it certainly took a differing path.[6]

1.2 Competing Theories

Like the character Beatrice "Tris" Prior from the 2014 movie, *Divergent*, somehow we took a divergent evolutionary path. As the scientific community came to appreciate the drawbacks of a large brain, so began a mad scramble to explain our divergent evolution. Before delving into how curiosity could hold the answer, it is worth reviewing the current theories on offer. While neither hold sway broadly across the scientific community, a few competing ideas do indeed exist and form part of the current scientific backdrop. The challenges of evolving a large, expensive brain have been well documented for a number of decades. As a result, various attempts have been made to explain what made our ancient ancestors different.

WHAT COMPETING THEORIES ON OUR UNIQUE EVOLUTION CURRENTLY EXIST?

One theory, posited by Colin Blakemore of Oxford University in England, is termed *macroevolution*. He believes that one ancestor experienced an extreme genetic mutation that randomly created exceptional intelligence compared with prior generations. Like a character from X-Men, this ultra-intelligent individual, Blakemore suggests, was so successful that he or she spread his or her DNA into the community at large through reproduction. Unfortunately, Blakemore's theory only goes back 200,000 years, attempting strictly to

COMPETING THEORIES

explain our recent intelligence rather than our multimillion-year journey since scampering around various woodlands. Our ancestral brain began to experience accelerated growth approximately 2.3 million years ago.[7]

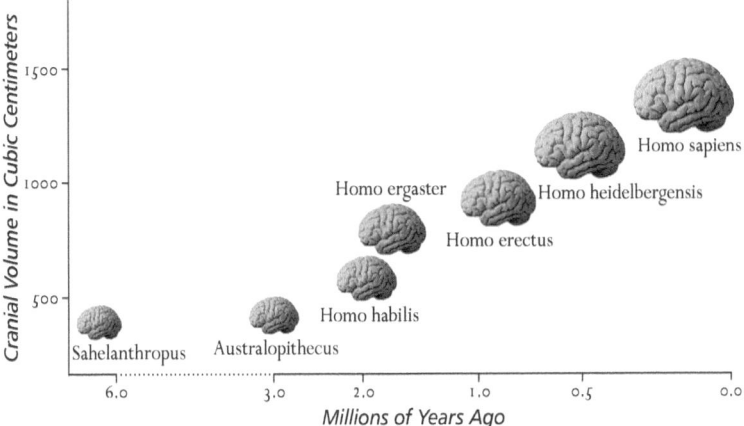

Based on the latest archaeological data, our ancestral brain capacity was approximately 400 cubic centimeters during the millennia before our divergence from the chimpanzee lineage over five million years ago. Around 2.5 million years ago, our ancestor was classed as Australopithecus and had a cranial capacity similar to that of the modern-day chimpanzee at 450 cubic centimeters, illustrating only modest encephalization during the prior 2.5 million years. Our Homo habilis ancestor of around two million years ago had a brain capacity of 680 cubic centimeters, up a staggering fifty percent compared with Australopithecus. A significant change had occurred during that interval. Homo ergaster, dating back around 1.7 million years, had a brain capacity of over 800 cubic centimeters, indicating continued fast-paced growth. Our subsequent ancestors, Homo erectus, Homo heidelbergensis, and Homo sapiens, had brain capacities of around 1,000, 1,200, and 1,350 cubic centimeters, respectively, further painting a picture of consistent encephalization during the past 2.3 million years. As a result, macroevolution seems an unlikely candidate to explain our divergent evolution.

Another theory involving nutrition is one of which Anthony Bourdain might be proud. Officially termed the *Expensive Tissue Hypothesis*, Richard Wrangham of Harvard University posits that cooked meat, in particular, resulted in more efficient nutrition intake, which, in turn, allowed our chimplike ancestors to have smaller, more compact intestinal systems that are better at feeding larger brains. Unfortunately, his theory does not preclude the possibility that improved nutrition resulted from intelligence and not the other way around; a classic chicken and egg challenge (or gut and brain in this case). Greater intelligence could equip our ancestors with various means of improving their nutritional intake thereafter, resulting in a circumstantial correlation between cooked food and brain evolution.

Furthermore, as with macroevolution, the timelines do not align well. Archaeological data does not support the widespread use of controlled fire dating back far enough to explain human evolution based on nutrition. The earliest unequivocal evidence of the controlled use of fire found at Israel's Qesem Cave dates back just 400,000 years. A more recent but isolated find at South Africa's Wonderwerk Cave dates back one million years, still shy of the 2.3 million-year target.[8]

Another challenge with the Expensive Tissue Hypothesis involves the many examples of creatures that get all the nutrition they require with relative ease and yet did not

COMPETING THEORIES

evolve greater intelligence. These include bears, walruses, and elephants to name a few. When members of a species get all the food they could possibly need and more, they tend to get bigger and bulkier but not smarter. In nature, fat is merely an efficient store of nutrients and calories. Throughout the eons, there have been countless species that dominated without evolving a large brain. When the window of opportunity arose to grow a larger brain, the opportunity was not seized as it was for our ancestors. An ingredient is likely still missing—excusing the gastronomic pun.

The Expensive Tissue Hypothesis places an emphasis on things getting easier for our ancestors, but natural selection works through mutation followed by pruning, where pruning involves the unfortunate reality of death and failure to reproduce for those lacking the beneficial mutation. In essence, evolution occurs due to conflict, friction, and hardship. An easing of challenges reduces the need for change and adaptation. The status quo could, therefore, endure, whereby the species would more likely prevail unchanged rather than become smarter. As Ralph Waldo Emerson wrote, consistency is the hobgoblin of little minds. He understood that comfort and safety do not encourage change or growth. Comfort and safety are likely the hobgoblins of little brains when it comes to evolution.[9]

Incidentally, the full passage from which Emerson's quote was taken, while written as a social commentary, could just as easily be a commentary on human evolution:

> "A foolish consistency is the hobgoblin of little minds, adored by little statesmen and philosophers and divines. With consistency a great soul has simply

nothing to do. He may as well concern himself with his shadow on the wall. Speak what you think now in hard words, and to-morrow speak what to-morrow thinks in hard words again, though it contradict every thing you said to-day. — 'Ah, so you shall be sure to be misunderstood.' — Is it so bad, then, to be misunderstood? Pythagoras was misunderstood, and Socrates, and Jesus, and Luther, and Copernicus, and Galileo, and Newton, and every pure and wise spirit that ever took flesh. To be great is to be misunderstood."

Ralph Waldo Emerson, *Self-Reliance*

Another leading theory on our divergent evolution is that an upright posture resulted in a larger brain. If this sounds somewhat familiar, it may be because of the long history of scientific research surrounding hominid evolution and bipedalism.

For over a century, scientists had believed the opposite to be true, namely that an upright posture was the result of a larger brain and not the other way around. This was even hypothesized by Charles Darwin himself who suggested that greater intelligence drove our ancestors to use their front paws more, meaning their hands, in exploration and to create tools or weapons. He believed that our ancient ancestors, therefore, shifted more weight to their hind legs due to this increased intelligence, resulting in an upright posture. Unfortunately, this was not quite how the story unfolded.

Given archaeological evidence not present in Darwin's day and modern dating techniques, we now know that the first stone tools appeared just over two million years ago, but our first bipedal ancestors date back fully four million years. Indeed, their brains were barely larger than those of today's chimpanzees. Our ancestors became reasonably upright well before greater intelligence emerged or the use of tools. Our ancestors essentially became upright apes long before becoming ultra-smart. It is now believed that an upright posture merely evolved for environmental reasons as it had in other (unintelligent) species over the eons. It is believed that the upright physical geometry of our ape-like ancestors was more conducive to life in drier environments as forests gave way to savannas.[10]

Despite the potentially circumstantial relationship between bipedalism and brain size, Dean Falk of Florida State University together with Christoph Zollikofer of the University of Zurich have put the egg firmly before the chicken, offering archaeological evidence that structural changes in skull orientation resulting from an upright posture may have, in turn, allowed a larger brain to evolve. Their theory, in essence, is that our hominid ancestors were bristling with potential to evolve a larger brain, but were simply held back by their status as quadrupeds. Unfortunately, it is widely accepted that an upright posture constricts the birth canal, thereby making a large brain more dangerous during childbirth and hence detrimental to natural selection.[11]

While it is true that bipedalism preceded accelerated brain growth, this does not automatically imply that bipedalism *caused* brain growth. There is no clear link between the two. Kangaroos, for instance, became bipedal with no bearing on brain size and indeed our ancestors were bipedal for over two million years before a significant acceleration in brain growth (or encephalization) took place. Why the incredibly long delay?

Darwin's theory on bipedalism remains relevant today despite the discrepancy in timelines. Intelligence would indeed necessitate the regular use of hands, resulting in a solidification or extenuation of an upright posture even if not initially causing it. As Darwin suggested, this would free the hands for activities other than a daily commute through the brush, but clearly an upright posture can occur without brain growth. Bipedalism is likely circumstantial in explaining the emergence of our intelligence. A smoking gun behind that intelligence remains elusive.

One of the longest-standing theories on human divergence is known as the *Social Brain Hypothesis*. Created by Robin Dunbar of Oxford University in 1992, he argued that the evolution of the human brain was driven by increasingly complex social relationships. The theory suggests that greater brainpower was required to keep track of individuals within the community, to know them as friend or foe. This, in turn, provided a survival edge through group dynamics of cooperativity. Over subsequent generations, as the theory goes, larger brains would repeatedly be selected. Essentially the brain evolved to be a living, organic Facebook *friending* system under this hypothesis.

Unfortunately, the theory does not explain why other species did not diverge in the same way. There are many examples of complex social interactions in nature, among both insect and mammalian species, without large brains appearing. Vampire bats, for example, keep track of friends. Research has shown that they deliberately select with which fellow bats to share food based on past interactions, shunning those who were previously less generous. Vampire bats achieve this social selectivity with tiny brains comprising less than two hundred million neurons compared to our 85 billion. Their brains are less than one-third of one percent of the size of ours. Much of that brain volume is dedicated to motor skills and olfactory processes such as radar and hearing, meaning that the number of neurons available to them for their own personal friending system is likely a far smaller fraction. Much of our brain mass is neocortex, available for varied tasks such as becoming a social media system, if necessary. In essence, vampire bats are pretty good at being social creatures with relatively minuscule brains. The search for a smoking gun continues.[12]

As with other theories on human brain evolution, Dunbar's theory has likewise not gained widespread acceptance across the scientific community. This is partly due to the recurring issue of what came first? The order of causation comes into question since it cannot easily be known whether complex social structures caused the brain to get larger or, in fact, social structures became more complex as we became more intelligent. Given the vast numbers of species that exhibit social behavior, it is likely that human social complexities and nuances are circumstantial to our intelligence, along with complex speech, iPads, and liquid soap.

We certainly benefit from greater neural processing power in keeping track of individuals within our social group. Cooperativity and group dynamics, in turn, aid survival, whether to hunt, gather food, stay in the vicinity of mates, or during conflicts. However, communal dynamics do not make us radically different to the many social species around us or explain how we diverged from them.

Theories on human evolution become stranger-sounding from there. Terence McKenna, an American ecologist and drug advocate, suggests that psilocybin mushrooms or "magic mushrooms," a naturally occurring hallucinogen and narcotic, took our ancestors "out of the animal mind" and into a new phase of brain evolution. Julian Jaynes postulated that mental aberrations such as schizophrenia resulted in greater mental awareness and even consciousness itself in his theory of the bicameral mind. Unfortunately, neither theory projects back to the earliest documented members of the genus Homo—namely Homo habilis—our hominid ancestors of 2.3 million years ago who had already, by that time, begun using stone tools.[13]

We are unique in the animal kingdom. How? Quite simply, the brain. It is our divergence, but is it our *primary* divergence or was another, more fundamental ingredient required before such intelligence could evolve? Perhaps there is an elusive ingredient previously unconsidered by the scientific community.

1.3 The Missing Link In Human Evolution

While our incredible intelligence represents our divergence from the remainder of the animal kingdom, it is not our primary divergence. Something came first. Missing from the picture is perhaps something inside us, termed *intrinsic*, rather than purely environmental. Environmental conditions change as groups migrate or long-term weather patterns change. Ice ages come and go every 100,000 years or so, for example. Access to food can change radically for a species from one millennium to the next.[14]

An intrinsic ingredient can drive the evolution of a larger brain on an on-going basis despite these variances and despite all the potential drawbacks of a behemoth skull-sack of blancmange on our shoulders, relative to our stature. Indeed, brain growth—or encephalization—occurred smoothly and consistently during the past 2.3 million-years despite all the inconsistencies in environmental conditions over those many hundreds of millennia. This is where curiosity comes in. Our *Curiosity Gene Hypothesis* aims to show that:

*CURIOSITY IS THE MISSING LINK IN
THE STORY OF HUMAN EVOLUTION.*

Evolution works through survival of the fittest. It involves random mutation followed by natural selection, but that is not the complete picture. For a feature to provide a survival benefit, it must be *used*. This fact is so obvious it typically goes without saying. When a creature evolves better eyesight, it provides a survival edge with no debating the extent to which that feature will be used since it is used automatically. The eyes are used immediately and unambiguously during various natural interactions. They are not kept shut, stating the obvious once more. The impact is direct and absolute. Likewise, a longer neck for eating leaves in the case of the giraffe or better camouflage for the stick insect both have direct and unambiguous benefits. In determining the evolutionary implications of a mutation, there is often no mention of an implementation step. It is typically too obvious to mention. Eyes are wide open.

Typically, but not always.

Consider the hand. A beneficial mutation would not necessarily stick around unless the creature had the mental capacity to use that improvement. Otherwise, the mutation would simply be a burden. Opposable thumbs get a lot of great press. They are indeed beneficial for smart creatures like those reading this text, but other creatures would only be rendered slower or weaker in having them. Digits working together in the same direction tend to be better for fighting or running on all fours. We had the mental capacity to grasp and use tools. Other creatures did not. Following any mutation, usage cannot be assumed or taken for granted. A beneficial mutation is only truly beneficial if it can and will be used.

There is a level of indirection in the deductive reasoning surrounding evolution, an extra step in the definition of natural selection, but one not lost on Darwin. He referred to it as adaptation, a loaded term that encompasses this extra step. A mutation had to be useful given all the variables involved including both the environment and the creature's mental faculties. Evolution fundamentally involves mutation, then implementation or usage, and then natural selection. The environment plays a role in the middle layer, hence the term adaptation. A mutation must be beneficial given the environmental conditions, but that having been satisfied, there is still the additional consideration of the creature's own ability to make use of the mutation. For speech and hand dexterity, we must ask this question, and the answer is intertwined in the brain's capability. We need the intelligence to take advantage of a potentially beneficial mutation of the hand or vocal cords.

So if the brain is needed to allow this wondrous array of distinctly human faculties to evolve, what about the brain itself? How did it get so powerful in the first place, especially given that other creatures, billions of species over millions of years, did not? What is the implementation or usage step in explaining a larger brain? How does a larger brain get applied? How does it impose itself on the environment to become relevant during adaptation?

The brain is not always used in the typical sense as might be a tool or appendage. When there are no immediate demands on an individual, that individual is responsible for pushing their own minds. However, the *individual* is the brain, meaning that the brain is responsible for pushing itself. Motivation is required. It is not the ability to use the brain that ensures it will be effectively used at all times, it is the *desire*. There must be a fundamental drive to use it, an internal impetus, one that has been with us for many generations.

Getting back to basics, for a moment, evolution works through random mutation followed by natural selection. Tiny changes in offspring from one generation to the next will eventually add up over multiple generations if beneficial to survival and reproduction. These tiny beneficial mutations are naturally selected and ultimately multiply. Now, consider a random mutation that increases brain capacity, meaning that an offspring is bestowed a marginally greater number of neurons. On the one hand, this provides the drawbacks of a larger brain mentioned earlier, such as infant mortality and dietary demands. On the other hand, does not a larger brain provide increased survival capabilities through intelligence? Unfortunately, it does not.

> *AN INCREASE IN BRAIN CAPACITY REPRESENTS ONLY AN INCREASE IN POTENTIAL INTELLIGENCE.*

An increase in intelligence is often assumed, but this is a false assumption. Intelligence is defined as the acquisition and use of information. The desire or impetus to make use of that increased brain power is the glue that ties brain mutation together with effective brain use, and hence survival benefits and natural selection. Stating the obvious, an unused mutation provides no survival or reproductive benefit and, therefore, would not stick around from generation to generation to become an embedded phenotypic attribute. Increased brain capacity through mutation requires a desire to acquire information and other drives associated with brain use to be beneficial. The desire to use our minds is the missing ingredient in the story of human evolution. We have it in abundance while other creatures do not. We are curious.

In software engineering, if we had the ability to build an artificial brain, curiosity would be a kind of app called a driver process or thread. As the name suggests, this is an app that drives the rest of the system, launching and orchestrating other apps or sub-systems. In the case of the brain, environmental conditions may occasionally provide mental challenges, creating external drives, but it is the continual internal drive of curiosity that tips the scales towards effective and consistent brain use. Otherwise, the brain's neurons would simply be underutilized, like idle CPUs in a computer system.

In the software industry today, Amazon Web Services, a subsidiary of Amazon.com Inc. founded by Jeff Bezos, has grown into a multibillion-dollar business by providing a set of pooled resources for other companies to leverage. Corporate customers, the likes of Pinterest, Airbnb, NASA, and Netflix, do not always need to purchase their own hardware. Instead, they may gain a cost-effective alternative through the shared hosting of computers at the Amazon site. These shared CPUs or chips can be driven at closer to full capacity compared to what could be achieved at an individual firm. At individual firms, sufficient computer power must be purchased to handle peak demand, but those same machines would remain relatively idle for large periods of time, off-peak, just as personal computers for home use are occasionally in standby mode as untapped resources. Amazon Web Services solves this problem, along with Google Cloud, Microsoft

Azure, and other cloud service providers. They simply loan out CPU power to one set of corporate clients during their busy periods while other clients might be experiencing lulls in demand. This results in economies of scale. Pooled computer resources avoid idling by being more consistently put to good use. Curiosity is the cloud compute service of the brain, loaning out mental cycles for other purposes when we are not busy struggling to stay alive in the face of immediate demands. Curiosity makes effective use of brain cells when they would otherwise be idle, underutilized, or in standby mode.[15]

Even within our species, we are not equally curious, and certainly not all curious in the same way. We manifest our drive for mental stimulation in a wide variety of ways, from scientific research to watching television shows, from reading to having conversations, from daydreaming to Facebook surfing, but to do nothing is typically unacceptable. We become infused with boredom, the negative sensation that drives us to act and to use our minds.

Other creatures occasionally exhibit signs of curiosity. Chimpanzees are among the most curious creatures in the world, perhaps second only to humans. However, research has shown a tremendous divide. While chimps have been documented lifting rocks to explore what lurks beneath without apparently seeking food, they will nonetheless spend countless hours lazing around in trees or on rocks. Humans have a very hard time remaining listless. We experience the discomfort of boredom that forces us to act.[16]

With a cerebral cortex of under 200 million neurons, less than three percent of the chimp capacity, dogs are considered less curious than chimps by researchers although strong emotional elements exist for their owners. We may be tempted to attribute their facial expressions with human thought processes, whether a raised eyebrow or inquisitive stare. Like chimps, however, dogs will also spend hours each day lazing around, conserving energy, curled up in a basket or on a rug. A truly curious creature, despite its inability to have a conversation, read a book, or assimilate a movie storyline, would find learning activities appropriate to its phylum whether rummaging in the undergrowth in its spare time, inspecting worms and insects, turning over objects, or generally leaving us wondering what on earth the animal is looking for. Indeed, alien observers might wonder what we are looking for as we scour our smartphones for mental stimulation during our morning commutes.

Human curiosity is on a different plane. We experience intrinsic curiosity, defined as a learning that has no clear and immediate relationship to a basic survival or reproductive need. Even if we spend time on our living room sofa, we still feel compelled to turn on the television, read a book, or thumb-stroke our smartphones. We must feed our minds or risk extreme discomfort and frustration. Indeed, we would appear psychologically abnormal if sitting quietly yet awake in a room without so much as a paragraph to read or movie to stream.

Any apparent curiosity we perceive of other creatures is quenched extremely quickly. A kitten will play with a ball of cotton. As well as a physical exercise, this may fairly be categorized a learning activity geared towards the acquisition of hunting skills. Unfortunately, the learning does not extend beyond immediate survival and hunting skills or information possessed by the parents. The curiosity is not intrinsic, as is normal for humans who strive to exceed the knowledge and skill levels possessed by parents.

Alison Gopnik, a child-development psychologist at the University of California, Berkeley highlights a fundamental difference between child's play and other mammalian play or exploration. Rather than simply acquiring basic survival and hunting skills, children, according to Gopnik, play by creating "hypothetical scenarios" with "artificial rules that test hypotheses," which in turn "makes children explorers of landscapes filled with competing possibilities."[17]

Gopnik associate and research scientist, Cristine Legare of the University of Texas at Austin, explains that children "explore causal connections through play." In one experiment, a four-year-old boy tested five different hypotheses in the span of two minutes. The experiment involved plastic blocks wired with electronics that would illuminate a platform when the blocks were arranged a particular way. The puzzle was to determine the correct arrangement to illuminate the device. Activation of the lights would depend upon both block ordering and orientation. The four-year-old, barely able to brush his own teeth or descend a flight of stairs using both feet as an adult would, conjures five hypotheses in attempting to explain how the blocks interconnect to illuminate the platform.[18]

For our ancient ancestors, curiosity meant that the extra neurons made available through mutation would tend to be used, not necessarily in the same way from individual to individual, but used nonetheless and with effect. This process implied a net survival benefit for our ancient ancestors. There were no televisions, books had yet to be invented, and the closest thing to Facebook was, well, a face, but our ancient information-gathering ancestors would experience a survival edge by learning about their environment, manipulating objects, discovering new sources of food, improving hunting or gathering techniques, being better prepared for changes in weather conditions, and more capable in battle against less curious counterparts.

In a nutshell, without curiosity, such a large brain as we have today would not have evolved. It would simply be a liability. Without an insatiable appetite to learn, stimulate our minds, explore, theorize, and test theories, we would still be primate, not human. The forces against maintaining a large brain are simply too great for a species to keep it around just in case it comes in handy. A small brain means a species remains relatively unintelligent but fast, efficient, and fertile.

To justify its own expense, a larger brain must be pushed into activity regularly and proactively. Neurons must be made to fire absent any urgent, pressing need. The brain must be put to good use without any clear and present danger, gathering information that will, on average if not always, prove useful at some point in the future. Otherwise, the burden of feeding the brain, carrying the brain, cooling the brain, immunizing the brain, oxygenating the brain, keeping it safe from trauma, and passing it through the birth canal would give the survival edge to the small-brained competitor...

WITHOUT CURIOSITY, WE WOULD STILL BE APES.

What makes us unique in the animal kingdom? Opposable thumbs? Speech? Creation and use of tools? These are all circumstantial to our intelligence, but even intelligence and brain size are themselves circumstantial to a drive, a desire, a thirst for knowledge. Predating our very intelligence is a core human impetus, an intrinsic motivation, an imperative to mentally excel. Curiosity is the most unique of human attributes, our defining characteristic, and our reason for knowing that we are truly special on planet earth.

PART TWO

THE ORIGIN OF CURIOSITY

2.1 An Elusive Drive: *Defining Curiosity*

So if curiosity is the missing link in human evolution, where did it come from? What made us divergent by acquiring this drive in our primal, ape-like form? This leads us to the next step in the investigation, and the next step in our *Curiosity Gene Hypothesis*: the origin of curiosity.

Before investigating the origin of something, it is important to know what exactly it is. In the case of curiosity, the definition can be refined to take into account its evolutionary impact.

> CURIOSITY IS A PREPARATORY
> INFORMATION-GATHERING DRIVE
> FOR POTENTIAL FUTURE SURVIVAL OR
> REPRODUCTIVE BENEFITS.

Defined by its preparatory nature, curiosity drives a *temporal* shift in behavior. Things are done ahead of time. Information gathered today is not necessarily used or useful immediately, but on average provides benefits at some point later in potentially unpredictable ways. Curiosity is a bit scattershot, in this sense. It takes advantage of prolonged downtime or lulls in the action, when we are not doing anything pressing, to create a somewhat artificial urgency today for what may come tomorrow. This urgency is part of an algorithm of curiosity. As mentioned, an algorithm is defined as a procedure or strategy for solving a problem or performing a task. The curiosity algorithm involves preemptive mental and corresponding physical activity. A time shift takes place whereby activity in the present without a clear and present need or danger nonetheless means being better equipped when the time comes, despite imperfect knowledge of what may come. We now have what software engineers would call a high-level functional design of curiosity.

Given this definition and as a quick, one-paragraph aside, we can perhaps shed new light on an age-old debate on whether nature or nurture is more fundamental in the formation of intelligence. Curiosity is, to a large extent, an inherited drive, innate in children and measurable from approximately the age of four. While it can change over the years, it certainly has a genetic signature. At the same time, it is a drive that impacts an individual throughout their life, imbuing intelligence on an ongoing basis. Information is, therefore, acquired rather than being naturally born. In the debate on whether nature or nurture determine intelligence, curiosity levels the playing field whereby anyone can seek greatness, nature and nurture working together for the betterment of self.

Clearly, curiosity drives information-gathering, but this is not to understate its purpose. For countless millennia, it has provided delayed benefits in an all-encompassing manner. This defines the algorithm or strategy of curiosity in a way that could be encoded into software and become part of a running computer system. Indeed, that already happened. Encoded by natural selection, the software exists in our heads.

Before pushing ahead with this new definition, it is worth looking at the existing definitions of curiosity put forth by the scientific community. A long history of research starting in the early 1900s has resulted in a dossier of information on the topic of curiosity. Unfortunately, given the lack of knowledge of computer systems during that era, much of this research created as many new questions as answers. Computer systems did not become mainstream until the latter half of the twentieth century, so understanding mental activity tended to be somewhat haphazard. The first stored-program computer was built in 1948 at the university that I, coincidentally, attended, the University of Manchester, England, by the renowned Alan Turing and others in a team led by Tom Kilburn.[19]

One classical attempt to explain curiosity involves the so-called *drive theory* whereby curiosity is seen as a naturally-occurring urge to be satisfied and satiated in a manner similar to how eating relieves hunger. While this sounds reasonable, researchers go on to apply classical psychology thereafter, from an era before the widespread adoption of computers and algorithms. In classical psychology, the focus was around *excitation* levels. We eat when excitation levels are high (meaning that we feel hunger) and have reduced excitation thereafter. Hence, we stop eating. Likewise for sex, with the obvious excitation level changes. When excited, we strive for sex and stop when excitation levels diminish. However, researchers were confused as to why curiosity was both triggered by high excitation levels and yet could result in *further* excitation rather than reduced levels. For example, as a discovery is made following an initial bout of curiosity, we often want to learn even more about the topic. This went against everything classical psychologists thought they knew about human behavior. Of course, algorithms have no such restrictions.[20]

Another approach to understanding curiosity in the last century involved *incongruity theory*. Here, volunteer subjects are noted as paying more attention to a topic when something is out of place, does not gel, or is in conflict. This piques interest and hence curiosity. While this is certainly hard to refute, researchers also noted that incongruity theory was inadequate as an explanation of all forms of curiosity. There are indeed many forms of curiosity that seek out knowledge, enlightenment, or wonderment without first having experienced inconsistencies.[21]

Researchers went on to define *trait curiosity* as something internal to explain the pursuit of hobbies such as learning to play the piano. Further subclassifications ensued, such as *breadth curiosity* and *depth curiosity* to describe an individual's range of interests (breadth) or degree of engagement in a single interest (depth). While appropriate for classifying a specific form of human drive, trait curiosity likewise suffered from the shortcoming of failing to explain all forms of the drive.

Exploration was also equated with curiosity. During studies conducted at the Max Planck Institute on the great tit songbird, a wanderlust gene was discovered, denoted DRD4, which appeared responsible for the birds having a greater tendency to venture further afield in search of food and other resources. Unfortunately, the study revolved around a bird, not necessarily a creature known for its desire to understand the universe, decipher the structure of a black hole, or learn how toenails grow. Hence, the implications for human curiosity remained questionable. In particular, it was difficult to discern the relationship between exploration and curiosity, especially given that cockroaches, flatworms, ants, and bees all have intense algorithms for exploration but could scarce be considered curious. Applying computer theory sheds more light on the topic.[22]

An algorithm, like a strategy, is defined as a process or set of rules to be followed in calculations or other problem-solving operations, often by a computer. The brain is a computer and indeed performs algorithms to solve problems, often unconsciously. In fact, any living creature possessing a set of neurons to generate activity are essentially running their own algorithms since they are solving the problem of staying in the game of natural selection. While humans hold the title for most neurons at 85 billion, even a fruit fly has an impressive 100,000 of them, an ant 250,000, and the repugnant cockroach a cool million. Each of these creatures, while not curious in any meaningful sense, are capable of executing complex algorithms including many that are exploration-related.

Clearly, the existence of algorithms or strategy alone does not imply curiosity. A smartphone map app runs sophisticated algorithms to determine directions, and a robot vacuum cleaner has its own algorithm to ensure the area under your bed does not look like Tunguska after the meteor hit, but these tools are hardly curious. Likewise, the great tit songbird and the not-so-great cockroach are unlikely to be exhibiting curiosity in their exploration antics. While curiosity clearly must have a genetic signature given that it is an inherited drive, the implications of the wanderlust gene, DRD4, on this trait remain questionable.

AN ELUSIVE DRIVE

The scientific community from 1900 to 1970 did extensive research on the motivational or drive aspects of curiosity, admirably, even going as far as to refer to it as an instinct. However, since then, curiosity research has almost exclusively revolved around *cognition theory*, where curiosity is viewed as a cognition, meaning a mental process of acquiring knowledge and understanding by thinking, experiencing, and sensing. This is perhaps an unfortunate shift. It somewhat undermines curiosity as a drive, an insatiable appetite that kicks us out of bed or makes us tick. As a species, we are *determined* to seek knowledge as if such determination were a basic instinct. Research that focuses on knowledge acquisition and subsequent thinking, while valuable, risks overlooking the urge that necessarily precedes cognition itself. Indeed, the depth and breadth of the cognition or information-gathering that takes place is greatly influenced by individual desire.

With all these differing ways of viewing curiosity, it is worth mentioning a thesis by David Deutsch, British physicist and professor at Oxford University's Center for Quantum Computation. In it, he discusses the essence of what a rational explanation should look like. In particular, a rational explanation should be difficult to alter. An ability to vary an explanation with ease, he posits, is a sign that an explanation is bad. Conversely, a significant difficulty in attempting to vary an explanation is an indicator that we should gain confidence in that explanation. In his words, "The search for hard-to-vary explanations is the source of all progress...the basic, regulating principal of the Enlightenment."[23]

Drive theory, incongruity theory, trait curiosity, exploration models, and cognitive theory each represent varying ways of deciphering curiosity. Each approach may be appropriate for some manifestations of curiosity, but each lacking in their own right until the next approach steps in as an ice hockey substitute to pick up the slack and get in the game, relieving the others temporarily. The classical theories and models represent easy-to-vary explanations, reducing our confidence in them.

Curiosity has been a source of confusion over the past century likely because it encompasses so many facets of human life and thought process. Curiosity can trigger scientific advancement and wonder one moment, and then celebrity gossip or rubbernecking the next. Curiosity is multifaceted.

Is there an explanation of curiosity that is hard to vary? When we speak of human curiosity as a defining drive, we are generally referring to a form of self-motivation called *intrinsic curiosity*. The definition of intrinsic curiosity is itself in need of clarification given its own journey from classical psychology. The official definition of intrinsic curiosity came from the postulated concept of an intrinsic *stimulation*. This, in turn, was defined as an activity for which there is no result beyond the satisfaction of carrying out the act itself, a task existing purely for its own sake. The act simply *is*. While there is a beauty or poetry in this definition, romantically capturing the spirit of an activity devoid of external motivation, capturing youthful inquisition without prompting by a teacher or hopeful parent, it is unfortunately fatally flawed.

The old definition of intrinsic curiosity from classical psychology is lacking since it neglects evolutionary forces. In particular, it overlooks benefits that are ethereally floating around in some unknown future. Any human drive that has

AN ELUSIVE DRIVE

no purpose is unlikely to stick around from one generation to the next during natural selection's intolerant and lethal pruning process. Indeed, a relatively new brand of psychology, called evolutionary psychology, is dedicated to such considerations. While not specifically focused on curiosity as a drive, this field goes beyond cognitive psychology by postulating mental modules, like smartphone apps inside our heads, that each evolved for specific purposes. There is a module for predator avoidance, one for food preferences, and another dedicated to alliance-formation strategies. Each can be seen as a separate app inside our heads, geared towards one aspect of ensuring we do well in the game of natural selection by sticking around in one piece and reproducing. As conceptualized by evolutionary psychologists, there are modules for childrearing, since reproduction does not end when a baby begins, modules for resource allocation, since food and other commodities tend to be limited, and even modules for lying and deception, since evolutionary benefits can arise from bending the truth. The list goes on from there.

In terms of intrinsic curiosity, informally we can see that it is driven in part by an individual's imagination and desire to decipher the world and beyond, but it is not without a result. Results are hidden, masked in some faraway future, but they exist. Hence, a subtly different but more accurate definition of intrinsic curiosity, as discussed, involves seeking knowledge for which there is no immediate survival or reproductive benefit. This definition allows for future benefits. Indeed, to have an evolutionary impact, the knowledge gained would, on average, tend to be useful at some point in the future. Otherwise, curiosity as a trait would not have stuck around through natural selection.[24]

The concept of information being useful, on average, at some point in the future, with a strong emphasis on the word *future*, is crucial to a deeper understanding of curiosity and speaks to Deutsch's "hard to vary" requirement. Indeed, this temporal element is extremely hard to vary without inconsistencies immediately arising, as classical psychology has illustrated over the course of a hundred years of study. There is a delay between the information being gathered and it being used. This delay is the very essence of curiosity and the reason that memory is fundamental to learning since it facilitates the passage of information into the future as part of this delay.

As an analogy, consider delayed gratification where some action in the present reaps future rewards. A well-known experiment involving four-year-olds remains, to this date, the single greatest predictor of future success despite the young age of the subjects. In Walter Mischel's famous test conducted at Stanford University in the 1960s, children who could resist the temptation to eat one marshmallow with the promise they would receive two in twenty minutes tended to be more successful later in life. While this phenomenon does not directly involve curiosity, it does involve the temporary abandonment of a primal drive, eating, in favor of an investment in the future. Curiosity is similar in principal, but instead of fighting a drive to eat a marshmallow, curiosity fights relaxation, contentment, and resting on one's laurels. It has become its own drive, geared towards future benefits, and must continually compete with the shorter-term drives represented by our primal needs of survival (including rest, which involves energy-conservation) and reproduction. Familiar conflicts and emotional challenges abound as boredom impacts work, chores, or duties.[25]

Applying Deutsch's litmus test, the new definition of curiosity, revolving around an intrinsic drive that provides delayed benefits, is difficult to vary, giving us confidence that the definition is correct or at least not obviously suspect. In modernity, it is hard to imagine a world without the benefits of science and technology that we enjoy today. These are advancements that followed prolonged mental activity by curious individuals starting in their youth, decades before their vocations or hobbies provided benefits. In archaic times, the discovery and use of tree resin as a natural fuel to allow ancient torches to burn longer undoubtedly came about accidentally through focused observation by curious prehuman ancestors. In prehistoric times as today, the temporal aspect of the curiosity algorithm was likely a key ingredient of each discovery or advancement. Indeed, it is difficult to express curiosity differently yet satisfy the various facets of this formidable trait.

The new definition was arrived at by focusing less on feelings, such as pleasure, excitation levels, or boredom-avoidance, but rather the evolutionary implications. While it is true that satisfaction, pleasure, and boredom-avoidance can and indeed do result during curiosity-oriented endeavors, these feelings are circumstantial to curiosity's purpose. Curiosity represents a temporal shift in the cognitive process. The ultimate purpose of intelligence is to make decisions and take actions that increase chances of survival and reproduction at some point in the future, whether or not we are consciously aware. Feelings are merely part of the process.

While most creatures make short-term decisions, acting on their drives for immediate benefits of food, water, or a mate, curiosity allows us to make longer-term decisions. Other creatures act upon what is in front of them, typically. Our gratification is delayed through research, pondering, exploration, and science which in turn provide a newly evolved form of emotional gratification. The more significant and relevant the information we gather, the greater the emotional gratification may be. In other words, the greater the chance of future benefits, the greater our interest tends to be piqued. Feelings are complimentary but secondary to the algorithm. Armed with this new definition of curiosity, as a temporal shift app inside our heads, it may be possible to seek out its origin.

2.2 Of Chimps And Men

Seeking out an origin of various human behaviors rather than physical characteristics is the topic of evolutionary psychology, as mentioned earlier. What differs in our investigation is that curiosity is a core drive rather than a particular behavior. It is not a single activity such as childrearing or predator avoidance. Curiosity is behind many and varied activities, hence a smoking gun or single reason for such a non-trivial app may be elusive. However, there is anthropological data that holds some important clues.

When we look at the archaeological and genetic evidence, two pivotal moments appear. First, five million years ago, our divergence from the modern-day ape took place shortly before the time of Australopithecus. Second, around 2.3 million years ago there began a period of accelerated brain growth and the first use of stone tools around the time of Homo habilis. Let's consider the first of these two pivotal periods.

> *COULD OUR DIVERGENCE FROM THE LINEAGE OF THE CHIMP REPRESENT A PIVOTAL POINT?*

We will soon see that the answer to this question is, No! While each of these two periods provides a potential trigger for the emergence of curiosity as a trait, we will see that the pivotal point was 2.3 million years ago. However, to set the stage and tell a more complete story, we shall start by taking the first of these, our divergence point from the chimpanzee lineage, and see exactly why curiosity likely did not originate at that time.

The chimpanzee is our closest living ancestor. Recent studies have shown that they are more genetically similar to humans than they are to one of their other cousins, gorillas. We share 99% of our DNA with them, but with a cerebral cortex four times larger we number in the billions today while chimp numbers dwindle to below 300,000 worldwide and dropping. Such is our success due to this cerebral difference.

Incidentally, bonobo apes, based on the latest DNA sequencing, split from the chimp lineage just two million years ago. They represent a more docile and peaceful variety of ape and we, humankind, also share a common ancestry with them five million years ago since chimps and bonobos were one and the same at that time. However, we more often refer to our shared ancestry with the chimp, not bonobo, due to chimps' higher intelligence, elevated aggression levels, and larger numbers. Bonobos occupy a relatively small region in the Congo Basin of central Africa. Chimps are considered more representative of the overall chimp-bonobo lineage, and humans arguably exhibit more chimp-like aggression levels in general.

Five million years ago, the planet was in the midst of a prolonged period of global cooling. Ancient rainforests that had provided reliable sources of fruit year-round began to give way to more open woodlands and savannas. With this change, winter fruit became a scarcity. As meat was not regularly available, this meant that a new source of winter food was needed. In an interesting dietary shift, the chimp-like creatures sustained themselves on root vegetables or tubers, food that up until that time was not eaten by large land dwellers. Only worms and certain rodents had sustained themselves on roots.

Here is where things could have become curious, so to speak. It is speculated that the most successful of these ape-like creatures did not simply wait for the winter before seeking out root vegetation, but rather would plan for the winter by identifying the locations of root clusters during the summer months either simply noting their location or physically carrying them home to be eaten at a later time. In doing so, a delay was introduced. Information was not immediately acted upon by eating what was simply available but instead a separation was created, a temporal distance between the current act of gathering information and the future act of using that data. Over time, it is possible that the act of seeking out food sources in advance of being hungry became a *desire* to do so. If such a desire became part of the mental makeup of the creature, could it conceivably grow to capture the essence of curiosity? Not likely.

For these ancient chimp-like ancestors, there was a clear survival benefit to gathering data on future sources of food. If this was an early form of curiosity, it was hardly intrinsic. The material benefit was indeed clear, providing sustenance. However, the distinction may only be a matter of degree. The greater the separation between an information-seeking act and the survival or reproductive benefit, the purer or more intrinsic the curiosity. Given our definition of curiosity as a temporal shift in decision-making through information-gathering, this becomes only a matter of time, literally. In seeking an origin of curiosity, we seek to satisfy the basic definition. A drive to seek out tubers could be a step in the right direction as a rudimentary incarnation of curiosity, but how likely is such an origin?

Other creatures, of course, gather food ahead of time. Squirrels gather nuts, bees manufacture honey, leafcutter ants cultivate fungus, and the list goes on. As a result, this candidate as an origin of curiosity has its limitations. While our ancestors may have developed a more sophisticated mechanism for gathering food ahead of time, it becomes less of a "smoking gun" in light of the multitude of other creatures that perform similar tasks.

While the period five million years ago provides a reasonable source for curiosity, there are a number of additional issues with the theory. First, brain growth only mildly accelerated following our divergence from the chimp. Various species of the Australopithecus genus, an example of which is the famous Lucy fossil, experienced only slight encephalization between five million years ago and 2.3 million years ago, evolving a brain only marginally larger than the modern-day chimpanzee. Second, and perhaps more profoundly, during those 2.7 million years, barely a single ape found their way out of Africa. Curiosity, if it began five million years ago to any extent, did not get far. Only with the advent of Homo habilis did our ancestors migrate large distances despite having almost three million years to do so in Australopithecus form.

Human curiosity takes us places, even within individual lifespans let alone millennia. A period of almost three million years confined to a single continent, while infinitely justifiable from an ecological point of view given resource considerations, does not fit the algorithm. Curiosity would drive a pack across large distances and perhaps across the globe as generations passed. Indeed, eventually it did but not until 2.3 million years ago, fully 2.7 million years after our divergence from the chimp. Only then did a whirlwind of migration suddenly take place. Every species of the Australopithecus genus, of which there are at least five (depending upon classification methods), remained in Africa and specifically eastern and southern Africa. Not a single Australopithecus fossil dating back more than 2.3 million years has, to this date, been unearthed outside this region.

Another reason to not put too much weight on the human-chimp divergence point is how circumstantial that point likely is. The reason for the separation of chimp ancestors from our ancestors at that time could be purely environmental and have no bearing on what made us truly different. In fact, similar divergences take place throughout the animal kingdom regularly. In the case of Australopithecus, at least five different species of this genus have been unearthed, only one of which represents our potential ancestor. The other four simply did not make it as far as modern-day chimpanzees by surviving until now. The mere fact that other species became extinct does not make our divergence point with them any less significant.

After our initial divergence from the chimp lineage five million years ago, fossil records for a period of almost a million years are sketchy likely due to high foliage levels in the region during that time. Leaf litter on forest floors renders fossilization far less likely, fossilization being a fragile and improbable process at the best of times. One ape genus, named Ardipithecus, was dated to around 4.4 million years ago, but it remains a matter of debate whether it represents a part of our lineage given the large gaps in the fossil record. Thereafter, from around 4.2 million years ago to 2.4 million years ago, multiple species of ape have been unearthed categorized by the genus Australopithecus, all from various parts of Africa, mainly in the south and east. In fact, the genus name is derived from the Latin word for southern, *australis*. The primary contender for being part of our lineage among these is the species Australopithecus garhi discovered in modern-day Ethiopia. The name "garhi" comes from the local word for "surprise," and indeed the timelines are a surprise. Despite being dated to just 2.4 million years ago, its brain is only slightly larger than that of a modern-day ape.

If each of the other varieties of Australopithecus had survived as the chimps did, we would perhaps be speculating about other divergence points at various times between 4.2 million and 2.4 million years ago. If chimps (and bonobos) were not the only survivors, we would see potentially dozens of common ancestors in Australopithecus form living today with differing divergence points. The survival of the chimpanzee represents a fluke; isolated and circumstantial. Other species could simply have been unlucky.

A discovery made in 2007 further illustrates this point. Starting around 1.9 million years ago was the time of Homo erectus, a relatively recent ancestor of ours believed to have evolved from Homo habilis. However, one Homo habilis skeleton was recently discovered that dates back a mere 1.5 million years, meaning that Homo habilis and Homo erectus both existed at the same time (in different locations). This does not mean that Homo erectus did not evolve from Homo habilis, as some scientists subsequently concluded before being corrected. It simply implies that one group of

Homo habilis evolved into the smarter, large-brained Homo erectus, while another group remained roughly unchanged in another region. The two distinct species of the genus Homo had their own divergence point around 1.9 million years ago. Obviously, unlike modern-day chimps, Homo habilis is no longer around to publicize this divergence point. Nonetheless, curiosity as a trait need not have appeared when we diverged from the chimp lineage since other divergences clearly took place as highlighted by a lesser-touted Homo habilis divergence.[26]

Seeking out tubers to address nutritional demands in the savanna seems an unlikely candidate for the emergence of curiosity, particularly when our divergence from the chimpanzee lineage is seen to be unimportant and given the number of other species that store food without the trait of general curiosity emerging. This brings us to the second pivotal point 2.3 million years ago. This era presents a far more likely candidate to have ushered in our curiosity drive, and an extremely violent one at that.

2.3 War: *Curiosity's Violent Birth*

Around 2.3 million years ago, there began a period of accelerated brain growth and the first use of stone tools in the form of choppers and scrapers. This was around the time that Homo habilis began to evolve from one of the many species within the Australopithecus genus. Rather than preparing for food shortages in the winter, a more likely origin of curiosity at that time paints a far more sinister picture.

EVIDENCE POINTS TO A VIOLENT ORIGIN OF CURIOSITY.

When a species becomes successful and is squeezed territorially, instead of competing chiefly with other species, whether chasing them as prey, ensuring they do not become prey themselves, and avoiding countless parasites and bacteria, such a species can find that its main enemy is itself. A process of war-like activity ensues, and not only for humans. Even creatures with brains far less advanced than our own have been known to form militias, launch scouting parties, perform blitzes, murder, rape, kidnap, and occasionally enslave. These include ants, felines, canines, birds, and fish.

Around 2.3 million years ago, our primal ancestors may have been in the midst of potentially the most sophisticated of such conflicts that the world had ever known. Half a dozen species of late Australopithecus squeezed by climate change or their own predominance into conflict may have formed the backdrop in which curiosity itself was born. Indeed, as we will soon see, one particular piece of evidence dating back fully 2.3 million years represents a likely "smoking gun."

Evolution often creates surprising complexity, even absent higher brain functions. In the case of ants, a mere insect yet one of the oldest and most prevalent species on the planet with a biomass second to none, there are surprising parallels with human crime and conflict. Ants have been known not only to go to war, but also to demonstrate the earliest example of slavery. Groups of soldier ants will raid a neighboring nest, killing any defenders, looting food, killing women and children, so to speak, perform suicide bombings in the form of toxic attacks, and even engaging in cannibalism.

In terms of slavery, raiding ants have been documented stealing unhatched younglings. When the stolen ants hatch, they go about their duties not knowing they are in a foreign nest. They clean, collect food, and care for the young in the colony even though they do not share the same DNA and hence have been exploited. To draw a loose analogy, just as a forced human captive may eventually accept his or her new home—a condition called Stockholm syndrome—so too will a kidnapped ant.[27]

Other cases of in-species conflict provide their own particular variety of atrocity. In mammals, infanticide occurs with alarming regularity in the cat kingdom. Adult felines will massacre unprotected litters not their own. This illustrates one particular evolutionary response to in-species conflict. While other species may nurture and safeguard the young of other members in order to promote the species or clan generally and thereby propagate the DNA of the group, some would instead take to the murder of newborns to promote their personal DNA above their kin. This behavior has been seen in domestic cats, lions, and leopards, but also outside of the feline kingdom in prairie dogs, bottlenose dolphins, langur monkeys, and even plankton. When ovicide is thrown into the mix, meaning the equivalent destruction of eggs, many species of bird, insect, and fish can be counted among the child killers.

Moving away from infanticide and ovicide, in-species conflict manifests more war-like parallels in wolves and chimps. In northeastern Minnesota, a 1998 study found that 43% of wolves not killed by humans were determined to be killed by other wolves. The data suggest that the killing was a consequence of territorial conflict given that the deaths chiefly occurred in border areas where territories meet. In one example, as published by Lucyan Mech, neighbors killed three members of a wolf pack while two others disappeared, presumed dead, allowing the aggressors to take over and occupy the territory of the defeated pack. Jane Goodall noted similar behavior in hyenas in 1986, the creatures having been spotted making incursions into "enemy" territories to attack neighbors.[28]

Genocide is not unique to humans. January 7, 1974 marked the first sightings of chimpanzees at war. Until that day, scientists thought that only humans deliberately sought out and killed members of their own species. Like a militia, a group of eight chimps in Gombe, Tanzania, embarked on a blitzkrieg. Setting out at dawn, they traveled south quietly yet briskly toward a neighboring group, communicating only in hushed tones and gestures. They happened upon a lone male and attacked. One held the victim down while five of the others began beating frantically, screaming with excitement. Heavyweight boxers on steroids, even chimps in captivity in relatively poor shape have been measured with five times the strength of professional athletes. In time, the attack was over and the raiders returned to their territory. The victim slowly raised himself, bleeding from dozens of gashes. Crushed and battered, he was mortally wounded. Researchers returned to the site on subsequent days, but he was never to be seen again.[29]

Since that pivotal sighting, scientists have witnessed brutal murders of males, females, and infants alike, culminating in the extinction of two entire ape communities through a form of war devoid of any codes of conduct. While this can be seen as either war or a primitive form of genocide, the human parallels do not end there. In 1976, two young females from one such failing group were seen integrated into the neighboring group that killed their mother. Parallels with human rape, kidnapping, trafficking, and Stockholm syndrome can be drawn.

As well as attacking or raiding, chimpanzees also exhibit military-style parallels in forming scouting parties. Normally boisterous, vocal creatures in group settings, they will fall silent as they congregate to patrol their territory, travelling single-file to minimize disturbances to foliage that could clue enemies as to their presence. They survey the area, occasionally stopping without a sound to look for signs of the enemy.

It should be noted that not all chimpanzee communities go to war. It is prevalent in the Kahama chimp community in Gombe, Tanzania, with kill rates approaching World War II standards, but elsewhere chimp on chimp fatalities are a rarity. Researchers now believe that the encroachment of humans on surrounding areas has squeezed the Gombe groups into territorial competition and that they were relatively peaceful prior to this. Nonetheless, chimp raiding in the Kahama community illustrates the potential and indeed capability for sophisticated in-species warfare in primates with brains even smaller than our Australopithecus ancestors.[30]

Of course, the ultimate reason for such hostility or defense is evolutionary. Given in-species competition and limited resources or territory, natural selection dictates that the struggle to propagate one's own DNA will culminate in direct confrontation, including murder. A pacifist alternative would be short-lived as more aggressive neighbors render the pacifists themselves victims, wiping out their DNA and resulting in the widespread propagation of the aggressive clan's DNA, causing the cycle to magnify and continue.

These are examples of conflict and war-like behavior in the animal kingdom. What of us? Human contemporary history includes no shortage of war, of course, but what of tribal conflict before civilization? Was the ancient savage not peaceful? Was not war the result of modernity spurred on by governments, national borders, and armed forces? It turns out that this is not the case.

Going back up to 300,000 years, the fossil records are relatively comprehensive. Date ranges extending into the millions present greater challenges as evidenced by the relatively fragmented records of early genera of Homininae such as Ardipithecus, Orrorin, and early forms of Australopithecus, but Neanderthals lived between 300,000 and 50,000 years ago. Scientists are thus able to not only recover multiple fossil samples but even DNA from within them. Modern gene sequencing techniques have shown that the Neanderthal lineage split from our own just 400,000 years ago, likely from the species Homo heidelbergensis.

Having split from the Neanderthal lineage, our ancestors later (in Homo sapiens form) came to wipe them out 50,000 years ago, as has been well publicized, putting an end to our distant cousins. However, recent genetic testing has shown that not all Neanderthals were killed. Most of us share genetic signatures with them dating back to that violent encounter, implying a diamond-shaped lineage pattern starting with divergence 400,000 years ago and then a coming back together or convergence 50,000 years ago as we crossbred. Essentially Neanderthals form a part of our ancestry, albeit in a diluted fashion. So while early Homo sapiens likely killed many Neanderthals, they also took time out to interbreed with them, most likely through conquest and rape. Indeed, all non-Africans around the world including those of us in Asia, Europe, the Middle East, and the Americas have between one and four percent Neanderthal ancestry dating back to that time, meaning six billion of us.[31]

While not our primary ancestor of 50,000 years ago, our DNA is nonetheless 99.88% identical to that of the Neanderthal having already been part of our extended family tree. We were already cousins before that confrontation. Given our similarity, what do we know about Neanderthal culture prior to their catastrophic encounter with Homo sapiens?

Digs have unearthed smashed and burnt Neanderthal remains in the vicinity of their own dwellings, indicating that they would kill and eat their own, likely neighboring and competing Neanderthal tribes, having them over for dinner so to speak. The charred remains indicate that the enemy carcasses were cooked and eaten on a regular basis.

Enter humans, also known as Homo sapiens *sapiens*, the repeated use of the word "sapiens," meaning wise, deliberately added by the scientific community to further subclassify us from the likes of Neanderthal who are now officially termed Homo sapiens *neanderthalensis*. Early evidence of war in Europe dates back to 5,000 BCE in Belgium. Professor Lawrence Keeley of the University of Illinois, Chicago excavated villages surrounded by deep, nine-foot ditches designed to protect ancient villagers from aggressors. Fences comprising wooden stakes, called palisades, lined the back of the ditches. By 1987, fifty enclosed sites had been discovered and with them, evidence of genocide. In Stuttgart, southwest Germany, an ancient mass grave was found that contained the remains of thirty-four men, women, and children killed by axe blows dating back five millennia.[32]

Crow Creek in South Dakota is the site of a mass grave dating back to the early 1300s. Before the time of the European colonies, even Native American tribes apparently unmarred by Western influence performed genocide. Archaeologists unearthed the remains of no less than five hundred men, women, and children who had been slaughtered and scalped a century and a half before Columbus had set foot in the Americas. Interestingly, as researchers enumerated the remains, they found a relatively low proportion of young female corpses, concluding that many of the women were subsumed by the aggressors as forced mates and slaves.[33]

A glimpse into life before modernity can be seen, not only through excavation, but also through the study of remote tribes in the rainforests of South America and New Guinea. Such tribes provide valuable clues into our ancient past. Without the influence of modern civilization and its armed forces, these tribes nonetheless take part in raid-based coalition warfare on a scale that, per capita, exceeds both World War I and II.

As of studies performed at the end of the twentieth century and documented by Richard Wrangham and Dale Peterson, the chance of being killed at the hands of another human while part of the Jivaro tribe of South America was sixty percent. This incredible number implies that, at the moment of an individual's birth, there was a sixty percent chance that he or she would die not of disease, famine, snakebite, an accident, or carnivore attack, but at the hands of a fellow human being. This is three times the death rate during either of the world wars, which topped out at twenty percent. When averaged out over the duration of the twentieth century, the 100 million individuals killed during both world wars represents a kill rate of just one percent—docile by tribal standards.[34]

The Yanomamo tribe, also in South America, and the Mae Enga tribe of New Guinea, exhibited relatively low kill rates at around forty percent. Nonetheless, life in those tribes would be twice as horrific as a continual world war. Among the most peaceful tribes studied, relatively speaking, was the Gebusi tribe in the lowland region of New Guinea. With a kill rate of fifteen percent, peaceful nonetheless implies a kill rate similar to that of either world war.

As is the case in the animal kingdom, a pacifist alternative would be short-lived in human tribal life as aggressive neighbors would take the upper hand through surprise attacks resulting in dominance and prevalence for the aggressors and their DNA. A similar dynamic was previously alluded to in the seventeenth-century writings of Thomas Hobbes, absent knowledge of genetic heredity. In his seminal book, *Leviathan*, he described the necessity of central authority as a means to avoid inevitable preemptive strikes against neighbors due to distrust and bilateral fear. The concept was even dubbed a Hobbesian trap in his honor, though later also termed a Schiller trap, catch 22 (lending its name to the 1961 war novel by Joseph Heller), and indeed simply an arms race. In game theory, the phenomenon is included in the idea of a Nash equilibrium, a strategy for winning a non-cooperate game. In the game of natural selection, concepts of distrust and bilateral fear include emotional signatures that accompany genetic imperatives. Regardless of the emotional elements, the natural selection implications dictate the behaviors that prevail over time. The winner gets to have their DNA stick around.[35]

From Neanderthals to prehistoric Belgians to precolonial Americans to tribal New Guineans, a common theme is apparent. How does this tremendous history of violence, war, and abuse pertain to human curiosity?

> *THROUGH A GREAT NEED CAME A GREAT ALGORITHM.*

An early form of curiosity could tip the balance in favor of one group of feuding primate versus another. Preparation in the form of mental activity, including information gathering, could have become a survival tool suited for in-species conflict. This would be akin to the benefit of seeking out root vegetation ahead of a difficult winter period but on a far more pronounced and varied scale. Curiosity-oriented endeavors in the present would reap future benefits, especially if those endeavors involved seeking naturally-occurring weapons...

Enter the smoking gun.

In 1949, archaeologists unearthed an incredible find that not only supports this hypothesis, it literally represents a highly divergent event. Dating back 2.3 million years, their pivotal discovery was of a mass grave of over fifty baboons whose skulls had been bludgeoned in the vicinity of our ancestor of the time, Australopithecus africanus. Given the size of the indentations in the fractured skulls and the discovery of antelope humerus bones nearby, archaeologist Raymond Dart concluded that our ancestors had, for the first time in terrestrial history, become weaponized.[36]

The weapons, while not manufactured but rather naturally occurring in the environment in the form antelope bones, illustrate that the use of weapons preceded man. Baboons can be formidable rivals. Today's baboons are noted for their ability to tease and provoke cheetahs in the wild as if for fun. The ability of our ancestors to systematically kill baboons for food likely implies utter dominance over, not only baboons, but also many other rivals. Suddenly capable of doing away with a vicious and normally unapproachable foe, Australopithecine numbers likely boomed. In general, as generations pass, a dominant species inevitably multiplies and begins to exhaust available resources or land. This leads inexorably to in-species conflict as the victors became their own primary competition for those limited resources.

The concept of violence being a driving force behind the rise of humanity, in a general sense, is not new. The theory was first proposed by Raymond Dart himself following the discovery of the baboon mass grave and published in the American Journal of Physical Anthropology. It became the inspiration behind the famous ape scene in the classic movie, *2001: A Space Odyssey*, written by Arthur C. Clarke and directed by Stanley Kubrick, wherein a humanoid primate wields a large bone, discovering weaponry for the first time. In the movie, the primate shares his innovation with the group, which subsequently overcomes a competing group, killing its leader by bashing his skull repeatedly. While a fictional depiction, the scenario is realistic based on the archaeological evidence of indented baboon skulls.[37]

Dart's message alongside Clarke's literary undertone is that humanity rose from those violent beginnings and our rise was inextricably tied to our viciousness. While this theory draws an incomplete relationship between conflict and intelligence, and hence failed to gain widespread acceptance in the scientific community, Dart's hypothesis, like the movie's underlying theme, does highlight the likely role of in-species conflict during the pivotal period that culminated in our exodus from Africa 2.3 million years ago and the transition of the genus Australopithecus to the genus Homo. As Robert Ardrey subsequently wrote in 1961 in his seminal book, African Genesis, "Territorial compulsion is more pervasive and more powerful than sex."[38]

Unfortunately, neither Dart nor Ardrey provided a compelling reason why this led to the evolution of a larger brain other than to suggest, as Ardrey wrote, "…the use of the weapon meant new and multiplying demands on the nervous system for the coordination of muscle and touch and sight. And so at last came the enlarged brain; so at last came man."

While poetic, this appears to be lacking an ingredient. Greater demands on the nervous system may result in better fighting machines through natural selection, but 85 billion neurons are difficult to justify. Certainly, the losers die and generally have a hard time reproducing after such a thing! So the smarter fighters with somewhat larger brains would see their DNA multiply, but the mental demands of fighting seem insufficient to justify our brain size.

Throughout nature, there are many examples of relatively large mammals that come to blows. These include bears, lions, gorillas, dogs, and elks to name a few. While these species are not necessarily systematically involved in group-based coalition violence, where one clan is attempting to oust another, individuals may nonetheless battle over a mate, food, or territory. Often this is the result of incidental contact, but such conflict will indeed have natural section implications since the loser of a skirmish will fail to obtain the mate or the life-sustaining resources. In many cases, the loser becomes injured and therefore less capable of surviving in the wild, adding to the natural selection significance. Despite many examples of sparring mammals and the obvious benefits of being a better fighter, no other creature evolved the added billions of neurons. While lacking a weapon, technique is nonetheless integral to animal skirmishes. To explain our divergent evolution, it would seem that the "coordination of muscle and touch and sight" is likely insufficient.

One notable example of in-species fighting involves giraffes. A giraffe comes to blows by swinging its long neck to impact the rival's torso, essentially using its head and neck as a weapon. The noteworthy aspect of this technique is that a larger brain would be more vulnerable during impact, and so minimizing the number of neurons required while still allowing the "coordination of muscle and touch and sight" is clearly advantageous in this case. For our ancient ancestors—as with any mammal—excess brain size could eventual result in poorer performance in battle given the reduced speed and stamina combined with increased fragility. Natural selection typically favors compact brains with specialized additional neurons for fighting numbering far less than dozens of billions of additional cells.

So while Ardrey's explanation is indeed compatible with natural selection on the surface, there remains an unresolved question of degree. What additional ingredient can justify the degree to which this particular form of Australopithecine in-species conflict can result in so many billions of additional neurons?

Without an additional factor, a slightly larger brain would likely suffice to aid our primordial ancestors during weaponized fights. General intelligence would not likely result. The multitude of forces against brain size would continue to play a fundamental role, not least of which is infant vulnerability (due to abbreviated gestational periods implied by larger brains) that would not have boded well in times of hominid war and distracted parents. Given the lack of large brains in nature despite long histories of conflict, albeit unweaponized, brain size would perhaps marginally increase for our feuding ancestors, but thereafter likely struggle to maintain its trajectory.

Before explaining how curiosity can complete the picture, it is worth exploring evidence of primal weapon use generally. The first sightings of today's chimpanzees fashioning spears came about in 2008 in Senegal. The chimps of the Fongoli region use spears to hunt galagos monkeys, also known as bushbabies. While not robust spears for use in combat or conducive as javelins, being relatively jagged and roughly the thickness of a human thumb, they are effective to kill the monkeys as they cower in tree cavities before their edible remains are subsequently fished out. This demonstrates the capacity for weaponry in primates with brains smaller than our Australopithecine ancestors of 2.3 million years ago, given further credence to the view of weapon-yielding bipedal primates.[39]

In terms of prehistory, wooden weaponry generally does not survive the transition to fossilized form. As a result, archaeologists have typically focused on easy-to-find stone-tipped spears, a relatively new and advanced technology when geological timelines are considered. Such weapons indeed fossilize well and date back as far as 500,000 years in our lineage. They comprise a tip made of chipped and sharpened stone tied to a shaft with fashioned twine from woven plants or animal sinew and glued with tree resin for added stability. Simple stone choppers, on the other hand, have been around for 2.3 million years and are ideally suited for shaving wood to a point. During the 1.8 million year gap, between the discovery of choppers and the invention of the stone-tipped spear, it seems unfathomable that fully wooden spears were not prevalent given how advanced are stone spears, requiring the sheathing of a carefully crafted tip painstakingly to a wooden shaft. It seems highly likely that wooden spears were in use for eons before the oldest known stone-tipped spear.

One possible source of indirect evidence of wooden spears could exist in bone damage patterns and, indeed, research performed by Geoff Smith of College University in London set out to detect this very phenomenon. Unfortunately, they proved that damage patterns from wooden spears are often difficult to discern from patterns produced posthumously through gnawing, marrow extraction, and even trampling by animals. While the picture is not rosy, one particular damage pattern could hold some hope, according to Smith. Puncture wounds on the scapulae or shoulder blade may one day provide evidence of manufactured wooden spears dating back to those early beginnings, but this remains an area for future archaeological research.[40]

As a quick, one-paragraph aside, it is worth mentioning the implications of Smith's findings on primeval life expectancies. We could perhaps offer a new ingredient to include when considering such metrics. Life expectancy numbers can be extremely misleading given that the scientific community has, historically, not prioritized breaking them down in meaningful ways for laypeople. While the life expectancy in Paleolithic times—meaning the Stone Age from around 2.5 million years ago to roughly ten thousand years ago—appears as just thirty-three years, this includes infant mortalities that skew the average greatly. Each infant death contributes a single-digit value to the calculation of averages. This gives the general public a misguided perception of how long adults would have lived after first reaching maturity. A far more useful statistic, but very infrequently provided by the scientific community, is the post-fifteen life expectancy, meaning the life expectancy for individuals after reaching age fifteen. During the same period, based on fossil findings, this was calculated to be fifty-four years, a staggering difference of sixty-four percent. This squashes any misleading notion that our primeval ancestors failed to live far beyond thirty—a number subsequently echoed in the popular media. Nonetheless, to whatever extent prehistoric adult life expectancies were relatively short, meaning fifty-four versus the sixties, seventies, or eighties of contemporary times (depending on the region and recent decade being measured), the impact of tribal warfare should not be overlooked, particular given that death by wooden spear is virtually undetectable in fossils. Indeed, given the extremely high rates of tribal killing uncovered by Wrangham and Peterson (and the corresponding implications for our extended ancestry), it would be logical to conclude that tribal war has greatly impacted prehistoric adult life expectancy averages in perhaps untraceable ways. As with aging tennis players, aging tribal war veterans would gradually become less successful in battle and more prone to (potentially catastrophic) failure, likely skewing the statistics further. We are left wondering how long our distant ancestors might have lived had peace been the norm.[41]

Returning to the story at hand, what evidence exists of in-species warfare during the Paleolithic era? Unfortunately, evidence of hominid conflict is scarce due to gaps in fossil records and the general improbability of fossilization. Nonetheless, the earliest evidence of cannibalism dates back the better part of a million years. This gruesome trait, a bell-weather of tribal conflict, while not dating back the 2.3 million years that marked the onset of accelerated brain growth and the baboon slaughter, continues to paint a picture of in-species violence extending deep into our terrestrial past.

The species in question is Homo antecessor, famous for being the first member of the Homo genus to reach northern Europe and whose remains have been found as far north as Norfolk, England. As a result of recent findings, it is now less auspiciously renowned for being the first of our ancestors evidenced as having neighbors over for dinner as part of the menu, hundreds of millennia before the time of Neanderthals. Examination of cut marks on various remains unearthed in Gran Dolina in central Spain dating back 780,000 years shows a distinctive pattern. Parallel cuts at the base of a skull in the location of the sternomastoid neck muscle imply a grisly reality. Dismemberment was deduced to be the likely goal. Given the absence of carnivore tooth marks, the species itself was deemed responsible for the hatcheting.[42]

Such findings help clarify the long history of in-species conflict that our ancestral line has been involved in, likely extending fully the 2.3 million years since the time of the baboon mass grave. Dart and Ardrey were likely onto something. Their instincts that in-species conflict was integral to our own evolution was likely justified, but the question of a missing ingredient to further explain our divergence remains. Perhaps during that time of weaponized war and amid new mental demands, there appeared a new drive.

Enter curiosity.

In the context of primate warfare 2.3 million years ago, the primary contender for seeding curiosity is the search for naturally occurring weapons and effectiveness in their use. Curiosity opens the door to the discovery of varied weaponry and continued curiosity lends itself to innovation in their use through dexterity and tactics. Seeking the better weapon involves preparation, potentially continually. The individuals or groups that were easily satisfied with their state of innovation could be at greater risk of decimation, allowing the more curious or driven group to become widespread and hence their DNA to become predominant. Think of it as the earliest example of research and development. It would have been the ultimate arms race, extending over countless generations.

Such an origin of curiosity, if true, satisfies the definition of curiosity in a most visceral way. Activity in the present reaps future benefits tied to survival and reproduction, not immediately and not directly, but eventually and with extreme prejudice, to coin a military phrase. In protecting one's family and friends, in seeking out those who would do harm, mental activity abounds. To paraphrase Ezekiel 25:17, with great vengeance and furious anger, curiosity may have been born. In light of ongoing rivalry and warlike danger over thousands of generations, natural selection favored individuals who prepared ahead of time.

One element of weapon-seeking is particularly relevant to curiosity: analysis. Each weapon must be assessed through critical thought and practice before it can best be leveraged. This analysis step is particularly telling and rare in the animal kingdom. While exploration is indeed one element of curiosity today, it is not a defining attribute. Rather it is contemplation—in the form of hypothesizing, critical thinking, and testing of hypotheses—that is very much fundamental to curiosity as we know it.

The discovery of pummeled baboon skulls suddenly opens up a series of possibilities with regard to Australopithecine thought processes. Seeking naturally occurring weapons, of which the implicated humerus bone is merely one, is not a gathering exercise as a squirrel gathers nuts. While weapon-seeking indeed involves exploration and gathering, it also requires an essential aspect of curiosity, contemplation, to determine the best bone or stick to use.

Given the possibility that weapons were not abundant for our primate ancestors of 2.3 million years ago and, furthermore, that group members could be vying for the best weapons from a limited available set, this only adds to the need for analysis in weapon selection. Echoes of scientific methodology and indeed childhood inquiry abound. To contrast this analysis with the child research of Gopnik and Legare, hypotheses, testing of hypotheses, and exploring causal connections, potentially through play, may all have occurred as part of wartime preparation for Australopithecus. Weapons would be assessed, perhaps tested, and likely used during play with group members. Parallels can therefore begin to be drawn between this wartime preparation and the scientific method as we define it today.

Innovation would not end with the acquisition of the best possible weapon. Effective use would be equally important. Research and development would likely extend to determining whether to jab or swing, use one or two hands, how to parry or shield, and the discovery of enemy vulnerabilities.

To provide added context, consider modern-day chimpanzees. There are a number of YouTube videos of apes in captivity attempting to use branches in aggression. See, for example, "Crazy Chimps Fighting at the LA Zoo." Weaponry is not typical for modern-day apes, whether in the wild or in captivity. The degree of dexterity with which modern-day chimpanzees, bonobos, or gorillas wield weapons is barely on a par with a three-year-old human child. Feel free to also search for YouTube videos using the search words, "Three-year-old drummer." While it may not be surprising to see a three-year-old surpass a chimp in hand dexterity given that we are human, after all, it helps illustrate how much room for improvement likely existed in weapons use for our Australopithecus ancestors. While more advanced than modern-day apes, our ancestors of 2.3 million years ago would have been unskilled by human standards. Curiosity-driven practice and play ahead of time would have provided a significant survival boost for one member over another during those early millennia of weapon use, motivating preparatory behavior.

As well as these technical skills, tactical know-how would be another area for potential innovation, including up-front thought processes surrounding sneak attacks, defensible positions, flanking maneuvers, safety in numbers, and use of high ground. In the same way that relatively primitive chimps today perform raids or embark on patrols, our more sophisticated and already upright Australopithecus ancestors would have performed similar operations. In defense, curiosity could provide added expertise geared towards being forewarned, guarding, protecting the young, or hiding as a last resort. In attack, curiosity could provide tactical advantages when considering terrain, flanking opportunities, elements of surprise, strength in numbers, stealing weapons from neighbors, or lying in wait to ambush.

Compared with technical skills, tactical ability is somewhat more difficult to exercise. Beneficial practice and play, if it existed, would require more participants and more intrusive rules of engagement—or game rules. On the other hand, upfront mental activity is more accessible. Contemplation by replaying events in one's mind following a brush with death or envisioning hypothetic scenarios would better prepare one individual for success over another. This introspection would further echo Gopnik and Legare's hypothesis and hypothesis-testing, perhaps providing the earliest example on Earth of this mental phenomenon.

For our ancient ancestors, introspection could have involved mentally visualizing enemies in a hypothetical backdrop. Beyond simply analysis, scenario-playing, as the term suggests, can be thought of as a form of intrinsic play. For our ancient ancestors, this could have been one of the earliest forms of creativity, writing a story in one's head, if indeed such thought processes began to emerge amid the added billions of neurons.

With the freedom within the thinker's mind to explore "landscapes filled with competing possibilities," to use Gopnik's words, the educational benefits of such inner play would be virtually boundless. Wartime natural selection benefits would likewise be substantial through tactical preparation. We will later see how posttraumatic stress disorder today could also be an echo of such an ancient intrinsic thought process.

For humans today in our relatively comfortable modern environments, curiosity is not typically associated with personal danger. However, during in-species warfare, contemplation would be a fundamental survival tool. With kill rates likely beyond those in World War I or II, curiosity spurred on by horror likely provided a survival edge and, over the generations, could evolve into an innate drive stamped onto the DNA of our ancestors—a curiosity gene.

Unlike the conflicts seen between other animals, our ancestors were motivated to act ahead of time and without an immediate external stimulus. Hence, we use the term *intrinsic motivation*, an internal drive to act on one's own initiative. This can also be seen to be the missing ingredient from Ardrey's explanation despite his and Dart's deep appreciation of the violent survival imperative of the time. The unique element of how our ancestors rose to the challenge of weaponized in-species conflict involved preparation ahead of any conflict. Rather than focusing only on the moment of conflict itself and therefore the "multiplying demands on the nervous system," as Ardrey put it, an intrinsic motivation would place demands on our ancestors' brains virtually continually, even when individuals could otherwise be resting and relaxing. These consistent mental demands magnify the significance of the pivotal archaeological discovery of 1949.

Preparatory behavior implies greatly magnified cerebral activity, justifying additional neurons. Rather than an intense but fleeting moment in battle, intrinsic motivation can result in added hours of daily mental activity, day after worrisome day, generation upon feuding generation. The natural selection implications of this consistent brain use are therefore skewed in favor of individuals with more neurons due to each neuron's magnified worth through repeated use.

An analogous effect exists in the technology industry. Software engineers have the potential to be paid quite well since their programs are repeatedly used. While lines of source code may take varying time and effort to develop, providing one potential measure of value for that software, it is by virtue of repeated execution that value is more fundamentally derived. When clients or users of a particular piece of software number in the hundreds of millions, that software tends to be executed millions of time each day. As a result, its value tends to be higher. When clients number in the thousands or less, the software tends to be executed less frequently and, therefore, tends to be less valuable, other things being equal. Likewise, added neurons appearing through random mutation can acquire magnified worth under the auspices of an intrinsic motivation, activating those neurons more frequently.

Of course, many factors contribute to determining the salaries of software engineers, but the concept of repeated use should not be underestimated. When an individual is paid to make a cup of tea—to provide a very British example—we can attempt to put a value on that work as a percentage of the revenue generated, venture capitalists notwithstanding. If the cup sells for five dollars, we could perhaps allocate one dollar for labor. When demand for a second cup of tea materializes, something astonishing happens—British sarcasm provided free of charge—the worker must now make another cup! Lo and behold, the effort expended in making the first cup of tea does not contribute to satisfying subsequent demand. The worker must start over in order to make that second dollar. Economies of scale can alleviate this problem when large batches of tea are brewed each time, but such efficiencies pale in comparison to those produced when software is executed billions of times during its lifespan. The work expended to initially create the software continues to reap rewards each time it is executed. Neurons in hominid brains likewise experienced cataclysmically multiplied worth when a disruptive new technology appeared, called curiosity.

In terms of timelines, the subsequent 650,000 years, from 2.3 to 1.65 million years ago, ushered in accelerated brain growth and signified the transition from the genus Australopithecus to the genus Homo, a term that literally translates to "man." Our ancestors for the first time left their adolescent cocoon of Africa, presumably possessing their recently acquired curiosity gene, finally venturing out to begin their conquest of other continents in a manner to which we have now become accustomed. Intelligence alone, without an adventurous drive, may not have resulted in such movement and, indeed, for the prior two million years did not. With a new and potent drive, multiple sects of the Homo genus spread across half of Eurasia. They reached as far north as Georgia bordering Russia, as far northeast as Beijing, China at the fringes of Inner Mongolia, and as far southeast as Java in modern-day Indonesia, a stone's throw from Australia. Along the way, they colonized India, Sri Lanka, and Vietnam. Such was the magnitude of the transition during those 650 millennia.[43]

As well as likely higher kill rates than those seen during modern warfare, primal in-species conflict suffers from the dreadfulness of being at home, at one's doorstep. Rather than protecting only fellow soldiers, each adult male must therefore also protect from slaughter or abduction their family members including mates and infants. Such a form of extreme warfare lasting hundreds of millennia, involving multiple clans and subspecies, would be relentless on multiple levels. A beneficial drive or impetus that might appear from such a challenge, given natural selection, would likely be passed down and magnified. Indeed, the very form of curiosity echoes this primal challenge, equally all-encompassing, equally unrelenting, and rarely affording individuals the opportunity to relax.

Unlike natural selection adaptations amid external challenges, in-species competition is self-perpetuating. From one generation to the next, the species improves through natural selection as the less capable become subsumed, but the bar would subsequently be raised. Challenges concerning the environment or other species are eventually surmounted, and balance is restored. This is not the case for in-species competition. When one tribe destroys another and takes over its territory, it grows as individuals reproduce. Over subsequent generations, the new community naturally fragments as groups become too large to stay whole and individuals lose track of their relatives. New groups are formed amid the advancement. These groups are essentially new tribes that then compete with each other in a virtually unending cycle.

Each future generation would be more advanced, having staved off prior competitors. Operating at higher levels of intelligence and sophistication, curiosity would continue to play its part. It would push one individual or group to excel beyond another or risk decimation. Even as the bar is raised and the stakes are upped, curiosity continues to eke out marginal benefits that can make the difference between life and death. When competition comes from within, each generation ushers in new and greater challenges. Curiosity, having been born, comes of age.

This potential origin of curiosity satisfies its basic definition involving a time-displacing algorithm: a drive toward information gathering in the present with a tendency towards future benefits. A violent origin also implies urgency and necessity. In-species conflict would represent a curve ball for an already well-evolved species such as Australopithecus, creating new demands. It would suddenly need to adapt but in a fluid and flexible fashion beyond anything seen in the past. From one conflict to the next, there would be no standard for success. Each prior success would merely up the ante for future interactions. Weapon choice, methods of use, and tactics would continue to make the difference between life and death despite prior advancements.

Australopithecus, relentlessly competing with its own kind after dominating others, suddenly had an exaggerated need to prepare in a contemplative manner. This, in turn, required effective use of available neurons. Any embellishment of neural capacity through random mutation would provide a survival edge like never before. Curiosity, analogous to an app, could emerge in the minds of our Australopithecus ancestors to satisfy its basic definition as an urge for time-displaced activity. A drive for preemptive exploration and analysis amid war came into being, which we now perceive as an intrinsic motivation. A curiosity gene became embedded in our DNA, a software blueprint of curiosity app version 1.0.

PART THREE

THE CURIOSITY GAME

3.1 Pretraumatic Stress: *Remnants of Curiosity's Violent Beginnings*

Curiosity is not all blood and guts, of course, but before investigating how curiosity may have changed from its violent beginnings, another pressing question is apparent:

> WHAT EVIDENCE EXISTS TODAY OF CURIOSITY'S INNER WAR?

The curiosity of individuals today rarely revolves around military benefits—rarely, that is, but not never. While curiosity may have taken on a life of its own, becoming virtuous in a sense, echoes of its violent past are nonetheless present today. When they erupt, they erupt vehemently.

Urgency is a common symptom of curiosity, particularly for those who personify the drive. The best scientists are both excited in their pursuits and fiercely determined. Rarely is the urgency of curiosity more apparent than during wartime or following disasters when mental activity becomes more pronounced as if an echo of its frantic origin.

World War II saw one of the highest invention rates that the world had ever known. During the mere six years of this intense conflict we invented radar, synthetic oils, synthetic rubber, the jet engine, rocket propulsion, nuclear power, nuclear warheads, microwave ovens, radio navigation (the precursor to GPS), and early computers in the form of the

Enigma encryption device and corresponding decoders made famous by Alan Turing and others.

The conflict of 2.3 million years ago lasted thousands of times longer than World War II. Such a conflict, lasting hundreds of millennia, barely fathomable by mere mortals such as ourselves who experience life on a daily scale, is on another plateau compared with a six-year war, as traumatic as World War II may have been. Hard to consciously grasp, it would be a seeming eternity involving countless generations of ancestors experiencing trauma, injury, bereavement, feuding, genocide, infanticide, rape, kidnapping, and blood-drenched revenge.

As higher life forms interact with their natural environment, emotional signatures are ever-present. It is worth remembering that emotions or feelings, in general, including happiness, love, pleasure, and pain, are not primary forces in natural selection but rather secondary to physical outcomes. They are part and parcel of the inner workings of minds, not to be mistaken for end goals in nature. This is often difficult for us to objectively comprehend since we are subjectively and unavoidably trapped within our own minds and guided by societal norms that emphasize feelings.

We may consciously think that we do things for feelings, but we evolved those feelings for specific purposes. Feelings-oriented apps are running within the subconscious parts of our brains. We may eat for pleasure and avoid the pain of hunger, both being feelings, but eating is a survival activity, notwithstanding the fact that society has now created an abundance of potentially unhealthy but delicious foods that we never had to deal with during our evolution. The ultimate purpose of eating is survival and not to make us feel better, which is merely the conscious perception of our interaction with the food.

We may believe that we mate for love and joy, but the goal is reproduction, notwithstanding the fact that contraception and abortion have likewise somewhat detached the activity from its original evolved purpose.

The list goes on. As well as romance and attraction, the reproductive imperative is also served by maternal and paternal instincts to ensure that the young are cared for and nurtured. These all include powerful emotional signatures that are secondary to the obvious evolutionary benefits. Thirst, fear, and shelter-seeking serve survival through self-preservation, each with feelings attached. Friendship, sharing, and cooperativity also serve evolutionary imperatives, since our survival includes group dynamics of helping one-another, but these again have emotional signatures that are secondary to the survival benefits.

Rarely is this distinction between feelings and true purposes more apparent than when the pleasure mechanism is abused by narcotics, a word derived from the Greek verb narkó, appropriately meaning "to make dumb." Narcotics are defined as any psychoactive compound with sleep-inducing properties but are typically associated with opioids, such as heroin. Opioids simply take advantage of evolved mental apparatus. It has been proven, for example, that heroin stimulates the same part of the brain that rewards humans for sexual activity, a reproductive imperative. These feelings were naturally selected to keep us involved in the genetic game of life from one generation to the next by encouraging us to reproduce, but opioids misguide this machinery.[44]

Another clear example of abuse of the pleasure and emotional mechanisms pertains to overeating. In nature, resources such as food were typically scarce and often expensive to acquire in terms of effort expenditure. If an opportunity to overeat appeared, individuals would rightly take advantage of it. Any fat an individual might gain would simply serve its purpose as an efficient temporary storage of calories and nutrients, tiding the individual over during lulls in available food. He or she would, nonetheless, have made efficient use of any fleeting overabundance, for example following a large kill. In extreme cases, in tribal times, a truly overweight individual would be inhibited in their ability to seek out and acquire more food if rendered slower, resulting in a natural equilibrium as the individual burned through their own fat reserves and eventually lost weight again, making him or her once again more nimble.

In today's society, food is typically abundant. As a result, greater conscious thought and deliberation are required to regulate food intake today compared with primal times during which our evolution took place and for which we have primarily adapted. Our attraction to sweet foods, likewise, was once conducive of a healthy diet since sweetness indicated the ripeness of fruit at a time when processed sugars and overabundance simply did not exist. As with detrimental drug use, detrimental food intake—including the excessive consumption of sweet, sugary foods—is also an abuse of our evolved machinery, highlighting the separation of feelings versus true purposes that modern society has now widened.

Feelings accompany activities. They are not bad, of course, but crucial to daily life. They motivate us and spur action, but the underlying reasons that such "feelings apps" evolved, becoming part of our mental makeup, are tied to the genetic imperatives of survival and reproduction. The problem is, these apps are hidden from our conscious perception. Human drive and motivation are orchestrated by feelings-oriented apps deep within our subconscious minds. They each evolved for evolutionary purposes, generating feelings from deep within to guide our conscious actions typically without our knowledge of those underlying purposes.

Curiosity is itself a drive. While relatively new in nature, it serves an evolutionary purpose. It improves our chances for survival and reproduction in the long term. As with other drives, curiosity likewise comes with potent emotional signatures. Following a traumatic experience or during times of war, anxiety spurs mental activity in search of answers and solutions. It is reasonable to envision our Australopithecus ancestors likewise going through prolonged and relentless periods of anxiety. Rising from the ashes of destruction, anxiety would accompany intense preparatory behavior during those early days of curiosity. In time, this anxiety—according to our *Curiosity Gene Hypothesis*—became part and parcel of an evolved trait towards such preparation.

Posttraumatic stress disorder or PTSD is defined as a mental health condition involving pervasive distress triggered by a terrifying event, whether experienced or witnessed. Examples of such terrifying events include war, terrorism, horrific accidents, and sexual or aggravated assault. While the "D" in PTSD stands for *disorder*, in primal times it may have been far from a disorder, perhaps even an asset, making the difference between life and death. Indeed, the symptoms of PTSD, as defined in the field of clinical psychiatry, read like a laundry list of survival techniques and best practices for primal warfare:

★ Persistently recalling or "reliving" events through intrusive flashbacks.
★ Recurring nightmares.
★ Distress responses when reminders are encountered such as similar settings or surroundings.
★ Avoidance of circumstances that trigger recollections of traumatic events.
★ Hyper-vigilance.
★ Difficulty in falling asleep or staying asleep.
★ Irritability or outbursts of anger.
★ Exaggerated startle responses.

Taking each of these, in primeval times reliving events and recurring nightmares can be practical, serving as preparation for future encounters through scenario recreation, whether consciously initiated or not. Symptoms of distress responses, avoiding similar circumstances, hyper-vigilance, and difficulty sleeping, while out of place in the comfort of modernity, would allow an individual to be forewarned ahead of potential recurrences or similar threats, facilitating risk-avoidance. Only in the safety of modern settings do such symptoms equate to a medical disorder. As Ray Bradbury said, "Insanity is relative. It depends on who has who locked in what cage." Indeed, an individual may be classified differently depending on their surroundings or community members. Finally, PTSD symptoms of irritability, anger, and startle responses, while shockingly irrational in modern civilization, can aid survivability in the wild through elevated energy, aggressiveness, and fast reaction times resulting in tenacious threat mitigation.[45]

In the modern age, where we are not in a constant state of exposure to the elements and enemies, the "D" in PTSD lives up to its name as being a *disorder*. PTSD may serve no practical purpose for a recruit returning from a tour of military duty or a trauma victim following an unfortunate but singular accident in the French Alps. For a permanent resident of the brush, however, with constant exposure to the elements, predators, and hostile neighbors, PTSD is practically PTS without the "D," meaning posttraumatic stress, a survival asset and scarcely a disorder.

Taking this chain of thought one step further, the terminology for such a condition need not center around past trauma but rather potential future traumas. The stress is preparatory in nature. As a result, and at the risk of seeming melodramatic or frivolous, we could assign a brand new word to certain forms of this anxious condition: *pretraumatic*. Of course, this is not to say that posttraumatic stress does not exist. It is simply worth emphasizing that our brains are great survival tools. Victims need not remain victims given the right preparatory behavior. In a prolonged wartime environment, PTSD could have served a practical purpose through constructive anxiety: veritably representing a pretraumatic stress.

This temporal element has a familiar signature. The anxiety, like curiosity, is preparatory in nature. There is a familiar time delay between thought processes and when those thoughts may one day prove useful. The anxiety of PTSD is perhaps a throwback to the earliest, most primitive form of curiosity, an anxiety in the face of trauma geared towards preparation, whether in protection or counterattack.

Another possible remnant of a cutthroat origin of human curiosity—as well as PTSD—is the tendency of some towards fatalism, prepping, or survivalism. These are individuals who invest significant resources into actively preparing for potential but improbable emergencies. Curiosity is itself a subconscious, deeply rooted preparation. It is feasible that the imperative for urgent planning and organization could be another echo of curiosity's brutal past. Prepping, while not curiosity in itself, requires constant questioning and simulation in the form of what-if scenarios and indeed "artificial rules that test hypotheses" as documented by Gopnik and Legare in their research on infant curiosity.

Prepping involves constructive preparation through agitated and anxious simulation of potentially nonexistent dangers. Examples of this feverish mental simulation include prepping for terror attacks, resource depletion, flash flooding, isolation, economic meltdown, pandemic, alien invasion, earthquake, mega quake, tsunami, nuclear reactor meltdown, nuclear war, chemical warfare, solar storms, cyber-attacks, prolonged power outages, tornados, landslides, wildfires, sinkholes, genetically modified food catastrophe, and mass poisoning. Here, preppers, rather than infants, become the "explorers of landscapes filled with competing possibilities."

In the early days of curiosity, if indeed it had a violent origin, preparation was a numbers game. The probability of one outcome over another in determining how to prepare and focus one's attention could make the difference between life and death. In modernity, it often makes the difference between lucidity and delusion as peppers are systematically considered part of society's erratic fringe, but modernity is, as the name suggests, a recent development that we did not evolve around.

A less extreme and far more common form of preparatory anxiety is simply environmental conservation and conscientiousness. Not typically deemed irrational, environmental conservation nonetheless involves a preparation for the future by avoiding shortsightedness and forestalling impending issues. Concerned citizens may be focused on the habitat of other species or future generations of our own. More expansive and less self-centered than prepping, pertaining to entire regions of the planet, distant peoples, or threatened species rather than self and family, conservation is nonetheless another possible echo of curiosity's origin. It includes a common emotional signature as that seen in preppers. In extreme cases, environmental conservation can even become militant. The Earth Liberation Front (ELF), for instance, founded in 1992 by members of the Earth First movement, made the FBI's list of top domestic terrorist threats having allegedly been behind various arson attack for their cause. The ELF resume includes a bombing of multimillion dollar homes in Seattle in 2009 deemed by them to have been built to subpar environmental standards.

While wartime inventiveness, prepping, conservation, and other forms of preparatory anxiety do not, strictly speaking, represent curiosity, they are nonetheless related. Curiosity is itself a form of preparatory stress (or pretraumatic stress).

Information gathered today can prove useful in the future. Information-gathering is an active endeavor involving exploration, hypothesis, testing of hypotheses, and persistent iterative thinking, building one thought upon another. If curiosity had a violent beginning, it would not be surprising that a potent form of this preparatory need in our prehuman ancestors remains with us today, reminding us from whence it came. A primal imperative, it once again expresses its trademark anxiety and rampant activity in preppers, PTSD sufferers, and environmental conservationists alike.

Another possible remnant of our violent primordial past is the phenomenon we call *sport*. While not directly related to curiosity, the concept of sport serves to reinforce the picture of consistent tribal warfare over an extended period of time by virtue of its likely origin.

Throughout the animal kingdom there are many examples of play, typically between younglings. As mention earlier, such play is geared towards the acquisition of immediate hunting and survival skills. As our ancestors acquired curiosity and used their minds to improve weaponry, it would not necessarily be at the expense of such play. The benefits of preparatory mental activity with regard to weapon innovation would not be at the expense of physical training and preparedness. Indeed, technical skills and tactical prowess would likely benefit from both upfront thought and upfront play, including group play. This implies both thinking and doing; both contemplation and practice. It is reasonable to expect play to have persisted and, indeed, matured as the brain encephalized. Play would embellish any benefits that preparatory mental activity would provide.

THE CURIOSITY GENE

In the field of evolutionary psychology, when a particular activity no longer serves a practical purpose but can be traced to an old survival imperative generations into the past, scientists refer to the activity or the desire for that activity as an *evolutionary side effect* (also termed a *spandrel*). Typical examples include sense of humor, enjoyment of music, and the ability to feel embarrassment. It is feasible that our compulsion towards sport is likewise a side effect, in this case, of primal warfare made permanent during innumerable generations of repetition.

Play—a close cousin of sport—would likely have been a significant part of preparation for conflict from a young age. As the conflicts became more significant and omnipresent, so too would the play or training. Over generations, creatures would evolve a desire for increasingly sophisticated games. As with other activities, natural selection would imbue pleasure and desire to ensure participation, compelling individuals to play. Those who trained would do better in battle and hence their genes would be selected, along with their play-oriented desires and pleasures.

As a corollary to our *Curiosity Gene Hypothesis*, we assert that this play eventually evolved into sports as we perceive them today, now devoid of much of the unsportsmanlike conduct that hominid warfare implies, such as murder and pillaging. The pleasure that drove preparatory play stuck around, genetically. As we became more intelligent, new games would be invented to titillate our pleasure mechanism, leveraging a larger set of neurons for physical training. It is therefore likely that sports are simply the inevitable remnant of 2.3 million years of primal warfare, continuing to satisfy our evolved thirst for physical preparation.

Of course, these physical elements are somewhat removed from the intellectual demands that likely led to the evolution of curiosity. Sport is the beefy, burly cousin of curiosity, the varsity linebacker to curiosity's nerd, tracing similar ancestry but with very different qualities and sensibilities.

While our desire for sports does not serve its original purpose as a preparation for conflict, it undoubtedly provides health benefits for those who actively participate in athletic or anaerobic sports. This is a fortunate consequence since the attraction we have to such competition is circumstantial to an ancient and, currently, absent survival need. On the other hand, for those who primarily watch sports without occasionally participating themselves athletically or anaerobically, the opposite could be said to be true. The stereotypical couch potato type who merely supports and cheers for his favorite team or athlete may be more likely to overeat and overdrink while watching excessive hours of sport per week. He is victim of a detrimental manifestation of this evolutionary side effect.

During times of tribal war, noncombatants had a tremendous stake in the outcome of a conflict. They supported their tribe, forming a part of their fanbase, in a sense. It would be saddening and perhaps even depressing to lose, given that lives would typically be lost and the group would be at greater risk of decimation or subjugation. These emotional signatures are familiar given soccer hooliganism and tearful scenes in stadiums around the world.

THE CURIOSITY GENE

We have become accustomed to seeing combatants in the form of professional tennis players breaking down in tears at the end of a Grand Slam final, whether as winners in relief or losers in dejection, but objectively this should strike us as strange. Professional athletes are simply doing their jobs. We typically do not see a Starbucks employee in tears of joy after making the perfect latte or a bartender sobbing into a poorly concocted martini. Just as athletes have a tremendous emotional stake in the outcome of a match, echoing the life-and-death struggle that seeded sports, so too will fans experience powerful emotional highs and lows that are otherwise hard to objectively justify. Lives are not at stake for either the athlete or the supporter, but they once were.

As with addiction to narcotics or unhealthy foods, modernity includes potentially detrimental elements when it comes to sports. In the case of passive supporters, access to television may satisfy a craving for sports without personal participation, hence increasing the likelihood of obesity and, potentially, heart disease, the single biggest killer in the developed world. According to data gathered in 2015 by the World Health Organization, of the leading causes of death in developed countries, including stroke, dementia-related diseases such as Alzheimer, lung cancer, colon cancer, kidney disease, and breast cancer, heart disease eclipses each of them by more than double.[46]

Another example of a detrimental side effect of sports involves the attraction we have towards games that, in themselves, are insufficiently athletic or anaerobic. Without mentioning specific sports so as to avoid offending sensibilities, such sports include relatively long periods of time spent waiting, whether standing or seated. Of course, this is notwithstanding preparatory and extremely beneficial gym time ahead of participation. By prioritizing physically demanding sports for both ourselves and our children—since early conditioning is a primary factor in driving such preferences—we set ourselves and our children up for futures with reduced risk of becoming victims to the biggest killer in the modern world, cardiovascular disease.

The existence of sports in their various forms and their importance in our daily lives continues to paint a picture of our long history immersed in in-species conflict. The primal benefits of preparatory physical activity in the form of play and the benefits of preparatory mental activity that seeded curiosity were not in conflict during those 2.3 million years. They were often complimentary as technical and tactical skills were honed and tested ahead of tribal conflicts. In today's society, where mental prowess provides far greater benefits that physical prowess for the vast majority of careers, sports remain as an evolutionary side effect to be kept in context. By ensuring that we spend more hours each week actively participating in sports than simply watching them, we can ensure the healthiest possible manifestation of this evolutionary side effect and primal remnant.

Incidentally, there is a gender-centric element here worth mentioning. In research performed by Richard Wrangham and Dale Peterson, and documented in their book, *Demonic Males*, tribal raiding was typically performed by males. As a result, it is perhaps not surprising that men tend to be more active in sports than women. According to the Bureau of Labor Statistics, 30% more men participate in sports on an average weekend than do women, and for a longer duration—two hours for men versus 1.4 hours for women. This serves to reinforce the likely evolutionary connection between sports and tribal conflict.[47]

Both mental as well as physical preparation were magnified during 2.3 million years of sophisticated in-species conflict. Of the two forms of upfront activity, it is the amplified use of neurons—seeding curiosity—that differentiates our evolutionary path most fundamentally from the rest of the animal kingdom. These were amplified intellectual demands most unique throughout nature resulting from a horrendous tribal war.

Certainty is often the best formula for success in nature. Creatures from phytoplankton to salmon to polar bears thrive on repetition, predictability, and staying within the norm. Other things being equal, their biomass grows with environmental regularity. However, when conditions change to rock the boat, being dependent upon a slow and steady course can spell doom. For instance, a change in freshwater conditions could have catastrophic effects on salmon spawning grounds or their convoluted trek upstream. Temperature changes can wreak havoc on the otherwise ferocious polar bear. Same-species conflict represents an extreme example of a dynamic and ever-changing challenge. The boat was truly rocked and remained so for eons.

In the case of humans, curious individuals are often repelled by certainty. This is yet another echo of our past. Boredom represents more than a mere discomfort for the intensely curious among us. It represents an anxiety that can become torturous, an existential dissatisfaction with life, a form of depression, a reason to rebel, to cry with frustration, and perhaps even to move to another country. Boredom and other emotions that render us restless evolved hand-in-hand with curiosity. If satisfaction and wonder are the carrots, then boredom and restlessness are the sticks. Curious individuals feel a rush, an anxiety, an urgency. They race to be constructive as if their very lives depend on it. Indeed, during a virtually endless war in a forgotten past, they once did.

In the woodlands of Africa 2.3 million years ago, wartime anxiety would have existed in droves for our primal ancestors. We share 99% of our DNA with the chimpanzee and the differences mainly revolve around their better sense of smell, their crouched posture, and their body hair. Only a tiny portion of that 1% difference dictates brain size. It is, therefore, safe to assume that Australopithecus likewise shared at least 99% of their DNA with us despite the lack of preserved DNA in their fossils to categorically prove this. Consequently, their emotional systems would have been virtually indiscernible from ours. During 2.3 million years of war, our ancestors were likely traumatized: *pretraumatized*.

In humans, our two amygdala represent our modules for processing and storing emotions. These almond-shaped components are located in the limbic system, an evolutionarily ancient part of the brain more primal than the cerebral cortex to which most of our brain size differences pertain. Given a sufficiently volatile and prolonged era of tribal conflict, one lasting hundreds of millennia, a new form of anxiety could evolve to prompt rampant mental activity. Rising from the ashes of omnipresent anxiety amid posttraumatic stress, a new drive may have been born.

Such an anxiety would be preparatory in nature, veritably a pretraumatic stress. It would be a recurring fretfulness and apprehension before a potential trauma rather than only after. It would find its way into our DNA, manifesting itself as curiosity as we know it today, a persistent preparatory anxiety. The curiosity drive endured even as in-species conflicts subsided, motivating us intrinsically as the ultimate safeguard of the future.

3.2 From War to Wonder: *The Demilitarization of Curiosity*

From its violent origin, curiosity somehow evolved into a source of wonder, scientific advancement, and exploration.

IF CURIOSITY BEGAN AS A TOOL FOR PRIMATE WAR, WHAT CHANGED?

Despite curiosity being widely considered a virtue, violence and conflict are likely needed for such a drive to initially appear since natural selection does not make meaningful changes without a genetic imperative.

Curiosity is one of our most defining attributes. Such a profound character trait is unlikely to evolve haphazardly. In the wild, a misguided trait could have mortal side effects. Energy and resources could easily be misdirected to the detriment of the individual or species. Natural selection would subsequently weed out such a trait. On the other hand, when things literally become a matter of life and death, a beneficial trait can appear.

When genetic propagation is at stake, nature may subsequently be forced to create a radical new urge through natural selection. Curiosity had a necessarily violent birth whereby those individuals or groups lacking the trait would be weeded out, such was the grave and lethal imperative of the time. Nonetheless, war-oriented curiosity somehow grew into a force for good, so to speak. It became demilitarized—a superhero returning from battle to help construct the community barn—but how?

Once embedded in the genetic makeup of Australopithecus, the curiosity gene could *later* perpetuate through traditional survival benefits beyond in-species conflict, ultimately resulting in less militant forms of curiosity thousands of generations later. This may come about for three reasons in particular: continuality, non-specificity, and opportunity.

Taking the first of these, war-oriented or militant curiosity is a *continual* drive. In the midst of primal conflict, every waking hour represented an opportunity to get a leg up on the competition and, conversely, every hour wasted a possible source of disaster through squandered learning opportunities. Once part of a creature's DNA, curiosity imbues contemplative energy potentially during any spare moments of the day. This means that the sheer impetus required for non-wartime benefits would indeed exist. The energy imbued by intrinsic motivation can be put to varied uses.

Second, war-oriented curiosity is *non-specific*. While it is in a specific sphere, namely a militant sphere, such a drive was never itself specific in its application. From the outset, it was oriented for adaptability, evolving to handle varied conflict dynamics in terms of weapons, interactions, tools, and strategies. This adaptability or non-specificity could eventually lend itself to benefits removed from in-species confrontation.

Third, there would be *opportunity* due to lulls in the action whereby in-species conflicts temporarily dissipated, particularly as groups of our ancestors ventured to the vast expanses beyond Africa. While in-species conflict could have been quite continual for many millennia for curiosity to appear, things would eventually change. Following the initial establishment of curiosity as a part of a creature's DNA, peaceful periods would, in time, occur. These represent opportunities for such a drive to seek non-militant outlets, particularly when combined with the drive's continual and non-specific nature.

To draw an analogy from the corporate world, continual innovation would help a company stay ahead of its competition. Corporate success would be the equivalent of survival. A research and development department, to that end, analogous to curiosity would provide the energy and activity backbone of that innovation. The company that prevails may eventually become dominant, perhaps even a monopoly, akin to a dominant species or dominant community within a species. In the corporate world, a monopoly may indeed reduce the incentive for innovation, but that is not to say that the company will immediately shut down and disband its research and development unit (or stop being curious, in our analogy). It may continue to flourish and produce new innovations even devoid of a survival imperative, perhaps having the liberty to then branch out into new spheres of innovation, new product lines, or new industries.

Humans have a virtual monopoly of planet earth, becoming dominant in the animal kingdom. We generally have an abundance of resources and in-species conflict is drastically reduced. We no longer, whether individually or as nations, feel compelled to fight to the extent seen in tribal communities. Our curiosity is therefore freed for other uses.

An initial demand on a species that is continual could result in a continual drive that does not easily dissipate as demands are subsequently reduced. DNA changes only gradually. Like the research department of a successful corporation, innovation endures for a time. Curiosity in our ancestors and ourselves has momentum, providing varied benefits long after the dust has settled on primate battlegrounds of old.

These three factors combined—continuality, non-specificity, and opportunity—suggest that it is possible for curiosity to find its way into spheres of life that are peaceful, practical, and indeed wondrous, even if born of violence. During lulls in conflict, it could have motivated our Australopithecus or early Homo ancestors to improve foraging, nest construction, hunting skills, hygiene, working cooperatively, or tackling social hierarchies. Once curiosity was embedded into their DNA, such an inherent drive could subsequently perpetuate due to non-militant benefits, finding its own momentum in the game of natural selection even devoid of violence, as the drive went from Incredible Hulk to Bruce Banner, from brute to scientist, an urge reinforced and embellished across the entire spectrum of life, embroidered in nature's mosaic.

Mark Twain once wrote, "Twenty years from now you will be more disappointed by the things you didn't do than by the ones you did do. So throw off the bowlines. Sail away from the safe harbor. Catch the trade winds in your sails. Explore. Dream. Discover."[48]

Rarely is this philosophy of adventure and risk-taking so faithfully illustrated than in the deeds of one modern-day innovator in particular, Elon Musk. Through his endeavors, he may be in the process of staving off the greatest threat facing our civilization this century, the exhaustion of fossil fuels and oil in particular. According to experts, there is a significant risk of a global economic meltdown far deeper and more prolonged even than that during the Great Depression of 1929 to 1939 and far more widespread if oil, a nonrenewable fuel, begins to run out given our current reliance on it. With seven billion people on the planet compared with a mere two billion during the great depression and hence many more mouths to feed, some projections, albeit controversial, infer the loss of billions of lives should we be forced into a subsistence lifestyle once more, particularly given our tremendous reliance on transported goods for sustenance.

In taking the electric car reputationally from a failed and laughable idea akin to an overgrown electric wheelchair to one of the most revered products on the market today, Musk has launched his own version of an arms race as potent as any involving Australopithecus humerus club caches or cold war nuclear arsenals but with a very different set of principles. Through Tesla, the car company of which he is CEO and chief product architect, and SolarCity, the solar energy company with which Tesla merged, Musk has begun to change the landscape of energy production and use, forcing established energy and auto companies to follow suit or risk extinction themselves. We depend, today, not only on fossil fuels for transportation but also for electricity since the vast majority of electric power comes from fossil fuels, 67% worldwide and over 70% in the United States.

Using similar mental machinery that helped countless ancestors survive tribal conflict, the highly curious of the world safeguard future generations. Curiosity-oriented endeavors provide tangible survival benefits through inventions from the plow to irrigation, from pottery to thatched huts, from the light bulb to X-rays. While a specific "smoking gun" may have been required for the curiosity gene to appear in the first place, once established, general survival benefits would then allow curiosity to flourish. Overcoming the initial hump in terms of the drawbacks of a large brain, the benefits of a large brain can ultimately snowball once a critical mass is reached, particularly when married with an intrinsic motivation.

A random evolutionary mutation producing increased number of neurons would be put to good use, allowing the human brain to continue to evolve even absent tribal conflict. Once the spark was ignited and assuming that curiosity became embedded into our ancestors' DNA, it is hard to imagine circumstances that would render it obsolete given that survival benefits tend to abound from this continuous, adaptable, and opportunistic drive. The same characteristics that benefitted weaponry form the very definition of the scientific method today, namely exploration, contemplation, hypothesis, and experimentation. Through advancement, they benefit all walks of life.

While it is hard to imagine curiosity failing to stick around, modern society, ironically, is posing the biggest threat that curiosity has faced in its 2.3 million-year lifespan. Due to modern medicine, birth control, and changes in societal norms, gene-selection from one generation to the next is undergoing a colossal change. It is difficult to predict what the future will hold. The prospect of genetically-modified humans becoming mainstream over the next few hundred years, while the fodder of sci-fi movies currently, is plausible, further rendering the future unpredictable. Nonetheless, patterns of natural selection for the vast majority of the past 2.3 million years have been relatively consistent up until today, allowing us to seek out patterns of evolution and decipher human origins. This, in turn, can help us better understand the human condition today.

Curiosity is not a hardware mechanism, akin to ant exploration, nor is it a typical software mechanism. It is a special kind of software. Curiosity acts as a driver app, a controlling software that pushes other apps into action, mobilizing varied parts of the supercomputer we call the brain. The curiosity driver app orchestrates activity. When other creatures are not tending to survival needs like feeding and staying safe, or reproductive needs like mating and rearing young, they typically conserve energy, lying dormant, relaxing until the next survival or reproductive imperative ignites. This is not true of humans. Our driver app is continually running during waking hours, triggering activity to fill the void or else punish us with uncomfortable emotions like boredom or frustration. While trapped on a crowded train or in traffic, we become agitated if mental outlets are not available. We are genetically programmed to seek out mental stimulation of varying sorts.

The curiosity app does not trigger a single, specific behavior. It triggers an array of behaviors that can themselves change over the eons. While some behaviors could become obsolete in time, curiosity remains relevant. Unlike a MapQuest map app, curiosity is not specialized. The Google map app may come along and render MapQuest obsolete, but curiosity remains relevant just as the phone's OS remains relevant. Even as the brain evolves to be larger and more sophisticated, the reward of curiosity does not diminish but rather grows with the brain, symbiotically, staying in harmony with brain version 2.0, continuing to tip the scales in a creature's favor amid varied survival challenges.

As wartime challenges subsided, curiosity sent its focus elsewhere allowing exploration, contemplation, hypothesis, and experimentation to thrive. Just as military technologies, initially closely guarded secrets, are eventually shared to benefit civilians, so too did the drive of curiosity itself. It evolved to become an all-purpose and all-encompassing civilian imperative. Indeed, as already touched upon, a similar effect can be seen in examining the military origins of certain technologies we take for granted today. The shortlist is impressive:

★ THE INTERNET: The underlying communication method, the Internet Protocol, was first used in a DARPA-sponsored network, DARPA being the Defense Advanced Research Projects Agency of the United States Department of Defense.
★ GPS: The Global Positioning System.
★ FREEZE DRYING: This was originally developed to preserve medical supplies without refrigeration during World War II.
★ MICROWAVE OVENS: The heating effect of microwave radiation was discovered incidentally in radar transmitters used during World War II.
★ PENICILLIN: While penicillin was discovered and named fourteen years before World War II by Alexander Fleming, it was formulated into a consumable drug, put forward for clinical trials, and ultimately mass-produced at the height of the war in 1941.

FROM WAR TO WONDER

- ★ VIABLE LONG DISTANCE AIR TRAVEL: The World War II invention of pressurized aircraft cabins allowed passengers and pilots to safely travel at high altitudes, taking advantage of thinner air and hence reduced wind resistance for aircraft.
- ★ JET AIRCRAFT: While Britain's Frank Whittle first took out a patent for a turbojet engine in 1930, it was not until the height of World War II that an operational jet engine was designed and tested by German engineers.
- ★ NYLON: Invented during World War II, nylon alleviated the otherwise excessive reliance on silk for parachutes.
- ★ CANNED FOOD.
- ★ THE WRISTWATCH: Originally used by officers, the wristwatch aided the timing of military operations.
- ★ UNMANNED DRONES: Now becoming popular in commercial filming and photography, and potentially soon for residential deliveries by Amazon and other firms, drones were originally military machines.
- ★ NUCLEAR POWER: The Manhattan Project of World War II succeeding in splitting the atom for the first time.
- ★ DIGITAL COMPUTING: The first decades of computing were almost entirely militarily funded including the so-called human brain, ENIAC (Electronic Numerical Integrator And Computer), used in 1946 to calculate artillery firing tables for the United States Army and, later, the Whirlwind, used in 1951 as part of the United States Air Force SAGE air defense system.

- ★ BEHAVIORAL SCIENCE: The social sciences of marketing and public relations originated from various wartime propaganda initiatives.
- ★ SPACE SATELLITES: The Soviet Union's Sputnik 1 was the first rocket to successfully achieve orbit, helping fuel the cold war arms race.
- ★ DIGITAL CAMERAS: These emerged from the photographic demands of space satellites for which conventional film was sub-optimal to say the least, not being reusable, requiring chemical development, and not lending itself to wireless image transmission.
- ★ ASTROPHYSICS, OCEANOGRAPHY, AND SEISMOLOGY: These disciplines each began as militarily funded endeavors.
- ★ FLY-BY-WIRE TECHNOLOGY.
- ★ DUCT TAPE.
- ★ JEEPS: The precursors to SUVs.
- ★ SONAR.
- ★ RADAR.
- ★ SUBMARINES.
- ★ AMBULANCES: These first made an appearance in the Spanish army during the 15th century in the form of "ambulancias," portable military hospitals that accompanied troops into conflict zones.

Just as many military technologies, initially beneficial during modern in-species conflict, became useful to all walks of life, it is reasonable to envision a similar metamorphosis with regard to curiosity. Once stamped onto a creature's DNA, it could become virtually self-sustaining, providing non-militant benefits and, therefore, continued natural selection of the drive.

As well as becoming general purpose, curiosity could also evolve over the generations in another fundamental way. The time horizon between the moment of information-gathering and its use could extend. Information gathered would not simply benefit a relatively near-term clash with rivals, but could instead benefit various longer-term needs due to the varying nature of those benefits. The temporal shift could grow in magnitude to produce purer forms of curiosity with benefits extending farther out. Individuals would understandably lose conscious sight of the potential survival benefits of their activities, partaking of activities for seemingly intrinsic emotional motives like satisfaction. Consciously unaware of the ultimate reason for their curiosity, natural selection would nonetheless reinforce the drive, rewarding such activities through the long, slow process of genetic propagation while leveraging emotions and feelings in the short term.

The greater the separation temporally between a curiosity-driven endeavor and the material benefit it provides, the more intrinsic and purer is that curiosity. For the individual, feelings associated with pure forms of curiosity do not necessarily revolve around survival. There is often no conscious appreciation of potential survival benefits when pursuing modern-day curiosity-oriented endeavors such as reading a book, having a penetrating conversation, or examining the news. When curiosity detaches itself from a conscious awareness of survival or reproductive benefits, associated feelings would necessarily lose their anxious edge and occasionally enter the domain of wonderment. With self-preservation no longer obviously at stake, other emotions begin to surround the drive.

Wonder is defined as the feeling of excitement, admiration, or amazement at something strange or surprising, sometimes in combination with awe. A laundry list of similar emotions and feelings include fascination, interest, intrigue, and immersion. While it is important to focus on natural selection mechanics, that is not to say that emotions do not come into play. Feelings and emotions indeed play a role, guiding actions. The subjective feeling of wonder is one such emotion. It evolved alongside the algorithm of curiosity to reward us with positive feelings. Wonder and its various sister sensations are the rewards for the curious. As Thomas Aquinas wrote, the fourteenth-century Dominican friar and philosopher, "Wonder is the desire for knowledge."[49]

When there are no wars to fight and as an individual's basic survival and romantic needs are met, the brain nonetheless takes flight, directing its energy at all manner of things, driving an individual to grasp the meanings of the stars, the sun, the wind, and moon. Eventually, there are CERN particle colliders surrounded by scientists seeking to boil down the meaning of life to a catalog of waves and particles with names like *boson* only to search afresh as new, unexpected questions arise.

During downtime, while others could be relaxing and reveling in their achievements of material success and comfort, curiosity-driven individuals are instead restless, dissatisfied, bored, or frustrated. These feelings are each subjective in nature but have motivational implications. Such individuals take to activities that others may see as a form of pain or unnecessary work. To the curious, such work represents less pain than the feelings of discomfort and frustration that precede such work, driving the individual into action and ultimately future satisfaction.

While negative emotions such as boredom may push individuals to act, there are also positive emotions that pull individuals. As with other human activities that have a genetic imperative, there are typically both emotional pushes and pulls, carrots and sticks. Positive feelings draw us in and provide emotional goals or rewards for our efforts. They are the carrots, each with slightly different hues and colors varying from interest to intrigue, from satisfaction to amazement, from immersion to obsession.

We move toward things that reward us with positive feelings, not just away from things that instill pain or discomfort. While we eat for the genetic imperative of survival, we move away from the pain of hunger and toward the pleasure and satisfaction of eating. The curiosity app makes use of similar emotional polar opposites to imbue activity. We are compelled by such feelings, often unaware of the genetic imperative the feelings are serving.

The feeling or emotion of awe is particularly interesting. It is associated with wonder and is defined as a reverential respect mixed with fear or admiration often of that which is grand, sublime, or powerful. Early connotations of the word revolved literally around the power to inspire fear or dread. The primary difference between awe and wonder is the degree of threat, power, intimidation, or fear involved. While awe and wonder are closely related, wonder typically does not involve these aggressive elements.

The close relationship between wonder and awe conjures images of curiosity's distant origin, linking its violent upbringing to its calmer come-of-age self. The word awe is often explicitly married with wonder in everyday terms, such as "awe and wonder," while other terms, such as "shock and awe," highlight its militant edge. Following the September 11 attacks and war in Afghanistan, the first phase of the attack on Iraq in 2003 was code-named "Shock and Awe," so named in homage of the military strategy of rapid dominance and overwhelming power. The term fell into disrepute following controversy over what some thought was excessive force. This became a political hot button, particularly following the publication by some European newspapers of the mocking headline, "Shock and Awful," but the word awe nonetheless illustrates a tremendous link between war and wonder, encapsulating both concepts at once. Awe is the reverence of the magical yet terrible.

From frantic pretraumatic effort mandated by primitive war many millennia ago, the emotional system evolved to not only provide anxiety to spur activity but also feelings of satisfaction to reward it. Awe can be seen as one step in the evolution of this reward, a force-inspired feeling of reverence. Over time, awe could become completely demilitarized, so to speak, holstering its weapon and retiring as peaceful wonder, intrigue, and interest. This chain of reasoning, while somewhat circumstantial in nature, does lend more evidence of curiosity's multimillion year journey from war to awe to wonderment.

Wonder and its sister emotions form part of the mosaic of human feelings, each with motivational implications, imbuing activity. Our brain, of which the emotional system is part, is guiding us as ever, optimizing our chances in the game of natural selection. We may no longer be at war with neighbors, but we are at war with fate, shifting the odds in our favor, impacting our destiny for the betterment of self and humankind.

Mental stimulation is rewarded with positive emotions in the present for what may come in the future. Such is the emotionality of curiosity, providing intrinsic motivation and spurring intelligence itself. Following 2.3 million years of evolution, we found ourselves not only anxious, frustrated, or restless to push us into preparatory action, but also interested, intrigued, and amused. From war, following a long and occasionally painful process of advancement, came wonder.

3.3 Why Only Us?

Only our ancestors evolved a curiosity drive despite four billion years of opportunity on earth since life began. Perhaps closely related nearby species also acquired curiosity whom we subsequently subsumed, but why only hominids? Given other examples of in-species conflict, why did curiosity not come about in many and varied species over the eons? Ants, wolves, chimpanzees, cats, and hyenas all participate in fatal in-species warfare, yet only our ancestors acquired the curiosity gene.

> *IF CURIOSITY ROSE FROM THE ASHES OF IN-SPECIES CONFLICT, WHY ONLY US?*

In identifying curiosity as the reason for the evolution of our intelligence, if indeed justified, we have deferred but not eradicated the question of how we became different in the animal kingdom. While we can assert that curiosity is a more fundamental and specific attribute than intelligence in defining the uniqueness of our species, it still begs the question, why did no other creature become curious? Quite simply, curiosity is an awkward evolutionary step.

The question of humanity's uniqueness now resides in a more meaningful domain involving the unique intrinsic motivation of curiosity. An understanding of the missing link in human evolution takes our inquiry to a more refined level. The next step in understanding our uniqueness is uncovering why the emergence of curiosity itself is not easy, and hence understandably rare throughout the animal kingdom.

For such a drive to emerge, a particular confluence of circumstances over a sustained period of time is required. Four elements, in particular, seem to stand out as prerequisites to curiosity's genesis. They are higher brain functions, contemplation, longevity (continuality), and variability (non-specificity). The last two may be familiar from the prior section, *From War to Wonder*, and this is to be expected. Some of the same attributes that made curiosity adaptable are also responsible for its rarity.

Taking the first of these, curiosity requires higher brain functions to take advantage of such a general-purpose tool. Since it is a driver app, a form of multipurpose software, it cannot run on an incompatible device. It requires a brain sufficiently endowed with neurons that can run the curiosity app and, in turn, allow it to effectively interface with other apps to perform varied functions on demand. In computer terms, this requires a complex system with multiple asynchronous components working in a disjointed yet harmonious fashion. It requires microservices, of sorts. Creatures such as ants evolved war-like behaviors without high-level thought processes as we use. Like robot vacuum cleaners, they became hard-wired to perform certain activities. An ant may also explore, but it would be using hardware mechanisms, meaning neurons, wired up for specialized functions. When curiosity makes us explore, it involves software interactions, neurons that are chemically configured through learning.

Throughout nature, benefits typically come through specific adaptations. There is often no need for a general-purpose adaptation. There is no need for a squirrel to evolve curiosity to gather nuts. It can leverage a specific behavior for nut gathering, meaning a hardware algorithm. Indeed, curiosity would likely be lost on the squirrel, not having the mental capacity to run such a piece of software in its brain. It could gather information until blue in the fur, but still fare better without experiencing the wonder of curiosity. The need for higher brain functions limits the set of possible acquirers of curiosity to a select set of mammals having hundreds of millions if not billions of neurons acquired through a slow and painstaking process of encephalization, or brain growth, without curiosity.

How, then, does a mammal acquire higher brain functions without curiosity, as did our Australopithecus ancestors and various other mammals? Malcolm MacIver, an engineer at Northwestern University, provides a plausible answer. He posits that when the first sea creatures ventured onto land 385 million years ago, meaning the primal ancestors of mammals, the resulting seventy-fold increase in the distance they could see meant an increased potential for planning. Underwater, light can only travel—or *attenuate*, to use the official term—a matter of meters. As a result, sea creatures travelling at speed must react almost instantaneously to predators and prey as they suddenly enter their field of vision with minimal time to spare. As sea creatures ventured onto land, they could see many miles, potentially. Not only did such creatures evolve the eyesight to take advantage of the light above water, but suddenly there was an evolutionary benefit to thinking for longer periods of time before making a decision or taking an action. The objects of potential future interactions could be seen many seconds or minutes ahead of time. Natural selection would reward a creature that evolved neurons for calculating potential scenarios or hypothetical futures. By effectively utilizing the added time to perform calculations and make better decisions, a survival edge would be attained. An early, basic form of planning would evolve due to the luxury of time that light afforded such creatures compared with life underwater. Mammals eventually evolved new thought processes and calculations that would not be beneficial to fish.[50]

If MacIver's theory on planning sounds reminiscent of the definition of curiosity as a temporal shift in decision-making, it is understandable. The same thought processes that made mammals successful paved the way for curiosity to appear. However, there remains a clear distinction. Intrinsic motivation involves thought processes not directly tied to a clear and present survival or reproductive imperative. When vision is required, a clear—and indeed *visible*—survival or reproductive interaction is often implied. Curiosity is fundamentally about going beyond the visible, beyond what is only in front of us or any given creature. Nonetheless, despite curiosity being on another plateau when it comes to contemplation, planning, and information-gathering, the implications of MacIver's theory when it comes to understanding nonhuman and prehuman mammalian brain capacity is extremely relevant.

Life on land and its implied visible vastness provided the context in which higher brain functions could evolve. Natural selection would reward creatures that had a somewhat greater number of neurons due to the benefits of increased calculation time, ultimately resulting in our Australopithecus ancestors, sporting around six billion neurons. As curiosity evolved, it leveraged these higher brain functions, providing intrinsic motivation to use them more consistently even when our ancestors could otherwise be resting and conserving energy. The driver app worked its magic, leveraging modules for planning and performing complex calculations to benefit vastly more distant hypothetical futures.

A creature with higher brain functions could be driven—intrinsically—to contemplation. The intricacies of preparing for warfare lend themselves to pure thought and analysis devoid of a visible need. While an environment on land supports a light attenuation that rewards seconds or minutes of calculation, circumstances were required that would reward hours, days, or even years of calculation. When Australopithecus sought naturally-occurring weapons, it would not be obvious what to gather. Each weapon would vary in size, shape, and material. Each would need to be assessed and potentially tested. This is quite unlike food gathering. A squirrel does not have this dilemma when gathering nuts, which are fairly straightforward in nature. There may be differences in quality from one nut to the next, such as ripeness or freshness, but these are easily assessed by leveraging specialized neurons in a hardware-equivalent form. Naturally occurring weapons, on the other hand, require particular attention.

By questioning the validity of each potential weapon, our ancestors began to contemplate, exhibiting this trademark aspect of curiosity. Each question of usefulness would cause a pause, a moment of reflection, a calculation of potential future benefits. The algorithm of curiosity, thus brewed in the melting pot of prehuman warfare, would be rewarded by natural selection for incorporating a significant degree of meditation and deliberation.

Animals classify the quality of foods using mechanisms that evolved over eons since the beginning of those foods, predating the act of storing foods. While squirrels, chipmunks, and hamsters all hoard food for later consumption, they select food using the same decision-making tools used in seeking food for immediate consumption. Determining the nutritional benefits of food due to variances in ripeness, quality, size, and type represents a relatively limited problem space. Furthermore, the same rules would apply from one day to the next, but weaponry creates unending variety and changeability, particularly if arms races occur, one tribe competing with the other for the best weapons or techniques.

Contemplation on the topic of effective weaponry may come easily to us today, but for Australopithecus, such preparation for conflict would have been an imperfect science. With barely six billion neurons compared to our 85 billion, constant room for improvement would have existed. The suitability of one weapon over another would be complex to discern. Clumsy by our standards, increased contemplation could make the difference between life and death. For our ancient ancestors, relaxing instead of contemplating could very well have been at their peril.

Temporally-displaced mental activity without a clear and present need is rare in nature. Thoughtful analysis in the present that can project itself into the far future is not the norm. Contemplation involves a pause, prolonged reflection, and a stillness to take things in. It is a period to mull things over and process information like never before seen in nature. With higher brain functions, such a transformation becomes feasible.

The second element required before curiosity can evolve, as well as higher brain functions, is longevity, meaning that circumstances must prevail for many millennia. Hundreds of generations are required before a new trait would become part of a creature's DNA. A temporary bout of warfare would be insufficient.

As rainforests in Africa slowly gave way to woodlands and savannas due to long-term but painstaking climate change, multiple species of the Australopithecus genus found themselves in relatively confined life-supporting regions for an extended period of time. Becoming the successful primate genus in Africa, as evidenced by bludgeoned baboon skulls, they would likely have been in fierce competition with one another for dozens of millennia. Given the difficulty in leaving Africa, due to its constrained exit point at the northeastern corner of the continent and arid terrain to get there, the stage was set for an epic.

What of modern-day chimpanzees that currently perform raids? Would they not become curious? As mentioned earlier and first postulated by Margaret Power in 1991, the chimpanzee raiding observed in Gombe, Tanzania is not representative of chimpanzee behavior generally but likely a recent occurrence due to reduced habitat caused by the encroachment of our own species. This not only explains why a small percentage of the groups studied participated in raiding, but also has implications for the duration of their war. It is unlikely that chimp raiding has had a sustained history over recent millennia.[51]

The third requirement for curiosity to emerge is variability. Only if demands change over time would there be a need for curiosity, otherwise a specialized, targeted mechanism akin to hardware would suffice as is the norm for natural selection. Flexibility comes at a cost. Curiosity is a flexible mechanism involving contemplation and analysis, but such analysis causes a delay. If a survival imperative requires a specialized skill, natural selection typically favors a less flexible but faster and more economical solution. As is true in software, complex calculations take longer to run. Large brains are often unnecessarily expensive solutions to problems in nature.

In-species competition provides variety on a number of fronts. As mentioned, there is a need for variety in technique since naturally occurring weapons exhibit unlimited variation in form. Additionally, the challenge was ever-escalating whereby the winning side would become predominant only to later fragment into new competing groups with advanced skills. Subsequent marginal improvements would ensure survival over and above the skills that were once sufficient to prevail. Generational improvements would also lead to ever-changing war strategy. Curiosity is an active endeavor. For it to be the best answer, changeability and variance likely pose the questions.

Together, higher brain functions, longevity, and variability represent strong obstacles to entry for would-be curious creatures, but could it happen? Is it possible that other species right under our noses are at this very moment becoming curious? While this is theoretically possible, human civilization, unfortunately, leaves no room for other creatures to capitalize on whatever curiosity they may acquire. They are likely to be thwarted in their efforts to either magnify their curiosity to potent levels or use it to become intelligent.

It took our ancestors multiple millennia both to acquire the curiosity drive and then to evolve a larger brain as a result of it. We are barely giving other creatures decades of a chance as we change the face of the planet from one generation to the next, let alone allowing hundreds of millennia to pass unimpeded. Indeed, species in the wild are becoming scarcer with each passing year as we are officially in the midst of the sixth mass extinction that the planet has experienced. As forests are torn down and endangered species are hunted around the globe, it is the first mass extinction in the planet's five billion year history caused by an organism—our species—rather than a natural phenomenon such as meteor, natural climate change, or seismic event as was the case for the prior five mass extinction events.

Species in captivity and pets undergo a process of domestication and managed breeding that distorts any potential for curiosity to compound if it were to appear in them, likely rendering the drive flaccid. Domestication is not kind to intelligence or the impetus to solve survival puzzles. Bowls of food and water are readily presented, and territorial battles generally involve which lamppost to choose rather than which neighbor to savage. A natural environment is required to allow, not surprisingly, natural selection. Smart or curious genes will not consistently be chosen from one generation to the next without it.

While this does not bode well for curiosity or intelligence emerging outside of humanity today, it is far more likely that curiosity was present in multiple parallel hominid species at various points over the past 2.3 million years. If so, they either failed to survive, we killed them, or we merged with them. The species Neanderthal (or Homo sapiens neanderthalensis) is one such example. We killed and crossbred with them. A distinct (and curious) species no longer exists on earth. Traces of Neanderthal DNA exist in six billion of us, but we are now *one*. All seven billion of us are considered Homo sapiens *sapiens*, regardless of these genetic traces.

Dutch scientist, Wil Roebroeks, found that, not pine sap, but the pitch of heated birch bark was used to create Neanderthal spear glue. This is officially the world's oldest known synthetic material and the first evidence on planet earth of an industrial process by any creature, and it was created by Neanderthal not man (meaning, not by Homo sapiens sapiens). It is clear that Neanderthals were not primitive and, in this sense at least, they were more advanced than our primary ancestors. Creating the pitch glue, similar in consistency to tar, involves a complex thermal process whereby the birch bark is heated to four hundred degrees centigrade, sufficient to condense the sap, but with safeguards to prevent it from burning. Scaled up to produce many ounces, this implies an industrial-like process that researchers have yet to emulate without the aid of modern tools. Indeed, there is little doubt that Neanderthals also had the curiosity gene, having evolved brains no smaller than ours, but what about before the time of Neanderthals?[52]

If there were species of the genus Australopithecus other than our ancestors that acquired curiosity around two million years ago, meaning that the genesis of curiosity happened more than once on earth, we are certainly the only species left standing. Only one lineage remains today. Other would-be sovereign rulers either fell on hard times or are simply lying in our wake. Those would-be rulers ultimately ceded the throne to us. If the king is dead, long live the king.

THE CURIOSITY GENE

There is no escaping our unusual place in nature. Abnormal is the new normal as humans are not typical of the animal kingdom. We conquered the earth, laying waste to forests and other natural habitats, impacting the climate, triggering daily species extinctions as we take our place on the organismic throne. However, we are the source of art, scientific advancement, community and comradery of unprecedented scales, charity like none other throughout nature, and the only species that attempts to understand itself and from whence it came.

Responsible for all that defines our uniqueness is our neural count and responsible for that is an app inside our heads. Without the curiosity app, our neurons would not have been allowed by natural selection to multiply to a whopping 85 billion and, without those neurons, we would not be capable of claiming the earth's throne or affording ourselves the luxury of charity, consideration, national and international cooperativity, or science and philosophy. With great power came great responsibility, to offer a cliché. With our tremendous brainpower came success and the luxury to be potentially generous to others, whether human neighbors or other species, despite whatever incidental imposition upon them we may have also placed.

While we may be hard on ourselves for what our success has done to other species, we also exhibit unprecedented generosity. We may debate whether it is sufficient, but it is certainly greater than the generosity offered by other species. Natural selection has simply not put them in a position to safeguard other species. They can ill afford such generosity, lacking either the resources or the brainpower. Despite our impact on the planet, we at least simultaneously strive to be mindful of other species in an unprecedented evolutionary paradigm.

Are we, therefore, good? The curiosity app was downloaded and installed from the iWar app store 2.3 million years ago. It was an extremely fortunate yet painful occurrence. It resulted in this array of wonder called humanity, all that is bad but also all that is good since, as Shakespeare wrote in *Hamlet*, "There is nothing good or bad but thinking makes it so." And we indeed think. Ravenously. Without such thought we would not be worth the classifying; we would not be worth the debate.

PART FOUR

CURIOSITY RISING

4.1 The Greatest Survival Tool On Earth

That was quite a journey for our species if indeed it is to be believed. Curiosity, the ultimate survival tool, resulted in the ultimate survival machine, humanity. The innately human drive of curiosity has been our chaperone for potentially 2.3 million years, holding our hands as we went from Australopithecus toddler with just six billion neurons or so, to Homo erectus adolescent, the clever journeyer, to erudite Homo sapiens, sporting 85 billion neurons. We became undisputed champions in the animal kingdom but did the journey end there?

> *CURIOSITY IS NOT ONLY BEHIND OUR EVOLUTION; IT CONTINUES TO SUSTAIN US.*

We now have a viable candidate for the origin of curiosity and how it led to the evolution of a larger brain. While modern gene sequencing techniques have yet to mature to the point of identifying human drives, the drive of curiosity is certainly inherited, along with others yet to be decoded. Given this multimillion year story culminating in this wonderful thing called *us*, what remnants of curiosity's importance remain today? Perhaps this: Curiosity is likely the single biggest driver of success, advancement, and

THE GREATEST SURVIVAL TOOL ON EARTH

intelligence on planet earth, continuing its role in allowing brain capacity to be leveraged to its potential. While this has obvious implications for individual careers and wealth, the basic repercussions on survival and natural selection continue to abound.

Extreme curiosity, occasionally appearing excessive and bordering on irresponsible, can save millions. In 1967, a mere stone's throw into the past, a single disease was responsible for the death of two million people. In today's age of worldwide panic over the Ebola virus in 2014 or the avian flu in 2003, neither of which took more than a few hundred lives, a figure of two million in a single year seems unfathomable. More shocking is that the same disease prior to 1967 had already taken 500 million lives since 1900, ten times the death toll during World War II, and countless millions more during the prior five thousand years since its first appearance. It has even been diagnosed in the mummified remains of various pharaohs in Egyptian tombs. Dubbed for centuries the scourge of mankind, the disease is none other than smallpox and a single individual, Edward Jenner, was responsible for developing the vaccine that eradicated it from the face of the earth, the curiosity of one man pitted against the killer of over half a billion.

In 1979, the World Health Organization finally declared smallpox an eradicated disease almost two hundred years after Jenner's discovery. At the World Health Assembly that year it was recommended that all countries cease vaccination: "The world and all its people have won freedom from smallpox, which was the most devastating disease sweeping in epidemic form through many countries since earliest times, leaving death, blindness and disfigurement in its wake."[53]

Jenner was more than an immunologist and more than a doctor. He was curious. During his early school years, he developed a strong interest in science and nature that continued throughout his life. He helped classify many species that Captain Cook brought back from his first voyage, even being invited by Cook to be part of his second voyage. He studied geology and carried out experiments on human blood. A year after the Montgolfier brothers first demonstrated the use of manned hot air and hydrogen balloons, Jenner built his own hydrogen balloon and flew it across the Gloucester countryside, a sight so unusual and otherworldly at the time that nearby farmers would not dare approach his landed vessel for fear of the unknown.

In 1788, Jenner became the first person to publish a paper on the parasitic behavior of the cuckoo bird, now known to be a *brood parasite*. A female cuckoo will lay a single egg in the nest of another species, the owner of which will unwittingly foster it, incubating the egg as if its own until hatched. The young cuckoo rewards the foster mother when she is not present by throwing her true chicks overboard to their deaths. So strange is this actual avian behavior that many naturalists in England erroneously dismissed his work as nonsense. For over a century, activists pointed to Jenner's cuckoo study to cast doubt on his vaccination work, slowing the widespread adoption of vaccination.

As a teenager, he once heard a dairymaid say, "I shall never have smallpox for I have had cowpox. I shall never have an ugly pockmarked face." It was a common belief that dairymaids were immune from smallpox, but it was Jenner's curiosity that ensured that the dairymaid's comments would stay with him. Jenner later concluded that cowpox not only protected against smallpox but also could be transmitted from one person to another as a deliberate mechanism of protection. On May 14, 1796, Jenner found a young dairymaid, Sarah Nelms, who had fresh cowpox lesions on her hands and transferred matter from those lesions to an eight-year-old boy, James Phipps. The boy fell ill with cowpox for two weeks. The following July, he was exposed to smallpox through an ancient and dangerous form of inoculation, implanting smallpox under the skin from whence it circulates more slowly rather than waiting until accidental inhalation. This means of inoculation, by exposing the individual to the live smallpox virus itself, while shocking by today's medical standards, had a two percent death rate, far better than the twenty to forty percent death rate normally associated with smallpox following accidental contraction. Before Jenner's invention of vaccination, it was the best alternative available. The boy had no reaction and, thus, it was concluded that he was indeed immune.

When Jenner wrote of his findings with excitement to the Royal Society, England's nineteenth century equivalent of the Department of Health, they simply rejected his findings. With annual death rates from smallpox at the time in Europe alone at 400,000 and a further 200,000 survivors each year being left blind, it would be a further three years before the vaccine would be accepted by authorities. Fortunately, however, the eighteenth century came to a close on a hopeful note with Thomas Jefferson himself being administered the vaccine in 1800, a century in which five kings had died of the disease.

Jenner, a multitalented and spiritual man with a flair for the violin and poetry, made no attempt to acquire wealth as a result of his discovery. Indeed, he devoted so much of his time and resources to vaccination that his private practice and personal life suffered, finding himself the subject of attacks and ridicule. Ultimately, his relentless and methodical approach changed the way that medicine was practiced. In 1801 he wrote, "It now becomes too manifest to admit of controversy, that the annihilation of the Small Pox, the most dreadful scourge of the human species, must be the final result of this practice."

While Jenner provides a breathtaking example of unwavering curiosity, he is not alone. Following his discovery, the microbiology behind diseases such as smallpox, despite vaccination, remained a mystery. Bacteria had yet to be discovered. By the middle of the nineteenth century, diseases such as cholera, anthrax, and rabies were not only a mystery but also the biggest killers on the planet. Though Jenner defeated smallpox and inaugurated vaccination, a term literally derived from the Latin word for cow, "vacca," it was Louis Pasteur who applied vaccination as a general technique.

Rather than relying on the existence of a similar disease or pox to trigger the body's immune system to fight a more deadly disease, Pasteur actively manufactured such microbes artificially. This removed the dependence upon naturally occurring benign or at least non-fatal equivalents of a fatal microbe. He infected rabbits with rabies and subsequently weakened the disease by drying the affected nerve tissue

THE GREATEST SURVIVAL TOOL ON EARTH

before administering it to humans. On July 6, 1885, a nine-year-old child, Joseph Meister, was mauled by a rabid dog and subsequently became the first human to receive the vaccine having little to lose. Despite the child's perilous fate following the attack, Pasteur could have faced serious consequences had Meister not survived the injections he was administering. The procedure was extremely controversial at the time, and Pasteur himself was, in fact, not a medical doctor. The boy survived, and a new era of medicine was ushered in. Today, the rabies vaccine alone is estimated to save over 250,000 lives each year since millions are potentially exposed to rabies through animal bites.[54]

Hailed the father of germ theory and microbiology, Pasteur inspired a new branch of scientific study. His work was not without obstacles. The prevailing opinion of the time and for the prior two thousand years since the writings of Aristotle was that food would spoil innately through a process called *spontaneous generation* and not due to the preexistence of living organisms. Prominent scientific figures publicly condemned Pasteur's views including, most vocally, Felix Pouchet, the director of the Rouen Natural History Museum in northern France. Undeterred, Pasteur experimentally proved that grape skin naturally contained living yeast that triggered wine fermentation. By extracting grape juice using a sterile syringe, he successfully inhibited fermentation, proving the existence of microorganisms and the falsehood of spontaneous generation.

Following his success with rabies, Pasteur went on to create the anthrax vaccine and establish the Pasteur Institute at which Waldemar Haffkine would create the vaccines for cholera and the bubonic plague. Pasteur also invented pasteurization, the technique named in dedication to him, extending the shelf life of milk which at the time was consumed raw resulting in frequent outbreaks of illness. Pasteurization is now used to help preserve countless foods including canned products, juices, wine, vinegar, syrups, and even bottled water, continuing to reap health benefits for millions around the world over a hundred years after his death. Pasteur's characteristic quote captures the scientific spirit and the essence of curiosity's temporality: "In the field of observation, chance favors only the prepared mind."

Bacteria, of course, is not the only danger we had to endure for millennia. Around the world, there are an estimated 400 million people afflicted with diabetes, 29 million in the United States alone. Each year, 1.5 million people die of the disease primarily due to poor healthcare in less developed countries. On average, it afflicts a stratospheric 9% of the population. Today, in developed countries, we accept the disease as a mere inconvenience to those unfortunate enough to have it but prior to 1921, diabetes was a certain death sentence. Only with the discovery of insulin by Frederick Banting, amid widespread criticism of his methods, was the death sentence repealed. Indeed, prior to his discovery, the British Medical Journal published a scathing critique of his experiments.

In the spring of 1921, Banting moved into a seven-foot by nine-foot apartment in Toronto with little more than an idea. Against the advice of his friends and colleagues, he left behind a medical practice to pursue research at the University of Toronto while having no research experience. Motivated by the death of a childhood friend, Jane, at age fourteen and in no small part by curiosity, the pivotal moment leading up to his discovery took place at 2:00 AM in the privacy of his home and within his own spinning mind. In his memoirs twenty years later he would write, "It was one of those nights when I was disturbed and could not sleep. I thought about the lecture and about the article and I thought about my miseries and how I would like to get out of debt and away from worry. Finally, about two in the morning after the lecture and the article had been chasing each other through my mind for some time, the idea occurred to me that by *experimental ligation of the duct and the subsequent degeneration of the pancreas, one might obtain the internal secretion.* I got up and wrote down the idea and spent most of the night thinking about it."[55]

Anxious mental activity in the middle of the night and, indeed, through the night is not foreign to the highly curious. Such "pretraumatic stress" is the very up-front mental work reaping future survival benefits that characterizes the algorithm of curiosity and its origin. Banting, being somewhat of an artist and writer himself, also offered these words on the topic of scientific discovery:

> "We do not know whence ideas come, but the importance of the idea in medical research cannot be overestimated. From the nature of things ideas do not come from prosperity, affluence, and contentment, but rather from the blackness of despair, not in the bright light of day, nor the footlights' glare but rather in the quiet, undisturbed hours of midnight,

or early morning, where one can be alone to think. These are the grandest hour of all, when the progress of research, when the hewn stones of scientific fact are turned over and over and fitted in so that the mosaic figure of truth, designed by Mother Nature, long go, be formed from the chaos."

Frederick Banting

While there is little doubt that scientists such as Jenner, Pasteur, Haffkine, and Banting were driven by personal inspiration rather than dreams of certain success and fame, having no guarantees in their endeavors, they did indeed achieve revered statuses despite multiple assurances by peers of their supposed follies along the way. Some scientists, however, fail to gain recognition in their lifetimes. Such was the case for the forefather of genetics, Gregor Mendel.

Mendel gained posthumous fame for his pea plant experiments conducted between 1856 and 1863 in which he established many of the rules of heredity, now referred to as the laws of Mendelian inheritance. Throughout his life, he was ridiculed by his peers for apparently wasting his time investigating the minutiae of minute plants, yet such is often the modus operandi of the intrinsically curious, gaining no clear and present reward. In Mendel's case, the reward was bestowed on future generations, his work having been discovered thirty years after his death.[56]

Curiosity, having evolved into an independent drive, necessitates no direct reward. We evolved emotions to provide motivation in the present, feelings that push individuals to leverage their curiosity, indirect rewards for what may come materially in the future. Beyond subjective emotions, the law of natural selection dictates that an average benefit emerges over time for either the individual or an appropriately sized group of genetically related individuals, regardless of any conscious awareness of this fact. In ancient times, the group size may have been a tribe of a few dozen. In modern times, the tribe could be billions strong. While curiosity provides a net material benefit, that is not to say that individual failures do not occur. Indeed, occasionally some meet perilous fates.

For daring to suggest the world was round, Cecco d'Ascoli, a Bologna professor, mathematician, astronomer, poet, and friend of Dante was burnt alive by the Catholic church in 1327. He became the first university professor to be executed by the Inquisition.

In the first half of the sixteenth century, a Spanish physician, Michael Servetus, discovered pulmonary circulation. He wrote of his discovery along with his controversial thoughts on religious doctrine that failed to gel with his scientifically oriented viewpoints. For his candor, he was forced to flee to avoid the Catholic Inquisition, working his way to Switzerland. Unfortunately, the local Protestant Inquisition were likewise not huge fans of his. They burned his writings at the stake, along with him that is, on the shore of Lake Geneva in 1553.

The Italian astronomer and physicist, Galileo Galilei, was a supporter of the Copernicus view that the sun is at the center of the solar system, not the earth. He was imprisoned by the Roman Catholic Church in 1633 for publicly contradicting the religiously inclined opinion of the time that the earth was at the center of the universe. Galileo said, "I do not feel obliged to believe that the same God who has endowed us with sense, reason, and intellect has intended us to abandon their use."

The discipline of science is rooted in curiosity in its purest intrinsic form. In this sense, one might call it a great virtue. While it will often be possible in hindsight to identify the benefits of a particular scientific discovery, it is often the intrinsic motivation of the individual scientist that is behind a given discovery. Activity is spurred by powerful emotions from specialized components inside the brain, and not any supposed material gain. Indeed, there is often no preconception of a material gain on the minds of the world's greatest discoverers. In Mendel's case, he was consistently ridiculed for his work. Galileo was rewarded with imprisonment and threats of torture. Thousands of others were killed.

Thirty years before Galileo's imprisonment, shortly after Copernicus released his pivotal work, *On the Revolutions of the Celestial Spheres*, Giordano Bruno took the work of Copernicus to another level in suggesting that stars were merely distant suns, each with their own orbiting planets, some of which could be life-bearing. A philosopher, mathematician, poet, and astrologer, he publicly dismissed the possibility that the vast and potentially infinite universe could have a religiously inspired center. For his attempted advancement of human thought, he was burned at the stake by the Roman Inquisition on February 17, 1600.

Until the time of the Enlightenment, the rainbow in its beauty was fervently heralded as a divinely conceived phenomenon. When the Roman scientist Marco Antonio de Dominis proved that it was produced by the solar spectrum in

1624, it did not go down too well with the authorities of the time. Dominis was systematically tortured and executed.

While there exist extreme examples such as these, most scientists, it should be said, muddle along without threats of imprisonment, torture, or their own personal barbecue by a picturesque Swiss lake. They simply find a sense of meaning in their ongoing pursuits without such drama. While today's scientists typically earn a living, the early scientists who established their respective fields often earned nothing. As the saying went, the love of learning and the love of money rarely met.

Other than the love of learning, it cannot easily be discerned what benefit was gained by the Greek philosopher, Aristarchus of Samos, back in 270 BCE. He was, in fact, the first to write of the earth orbiting the sun, not Copernicus or Galileo. Likewise, it cannot be conceived what external, material goals drove Heraclides of Pontus, a hundred years earlier, to cite the earth's rotation about its own axis in explaining the daily cycle for the first time in recorded history. These tidbits of knowledge were lost to humanity until Copernicus popularized them an unfathomable two millennia later, the knowledge having been inexcusably lost to the world.

Einstein once said, "I have no special talent. I am only passionately curious." In his greatness, he revered, not his intellect, but his drive to attain it. He was not alone. These examples illustrate an intensity that occasionally borders on a disorder, where individuals make sacrifices of sleep, comfort, socializing, entertainment, and even love—delaying gratification monumentally. While potentially driven by a traumatic experience or a great need, illustrating post-traumatic elements, the drive is very much geared towards the future, a personal drive that can often best be described pretraumatic. Stress and anxiety in the present can often reap rewards that extend beyond individual livelihoods or lifetimes. Curiosity is an intrinsic motivation for the eventual betterment of self, tribe, and humankind.

4.2 The Unreasonable Minority: *Evolutionarily Stable Strategies*

One thing that emerges from these examples of world-altering discoveries is that a precious few can impact millions.

> THE NEEDS OF THE MANY ARE SERVED
> BY THE DEEDS OF THE FEW.

The extremely curious are the salient minority, heralding humanity, ushering in a new species and safeguarding it thereafter. However, curiosity is not without its drawbacks.

For a society or group to be successful, it is not necessary for all individuals to be ravenously curious. Nature has a wonderful way of creating beneficial variety within a single species or tribe to, on average, improve group survival chances. Natural selection is not only about individual genetic propagation but also that of the group, clan, or tribe. When tribal communities are involved, the unit of selection is both individual and group-based. This is perhaps why curiosity is so misunderstood and often underappreciated. If all individuals had their head in the clouds, so to speak, the tribe might not survive for long.

Exploration and experimentation present risks. First, curious individuals invest significant resources in exercising their curiosity, whether Australopithecus or human. Calorific energy is expended. Furthermore, such individuals expose themselves to personal dangers of unusual foods, bacteria, viruses, snake bites, insect bites, bruises, sprains, fractures, and making themselves generally more available to predators and the elements. Those risks are often mathematically secondary to the benefits of curiosity, particularly during an era of prehistoric conflict when the risk of being slaughtered can eclipse even the dangers of the world wars, let alone snake bites. As mentioned, the manslaughter mortality rate measured in South American tribes, a bellwether of our extended history, tops out at fully sixty percent, four times that of either world war. Curiosity provides magnified survival advantages during times of conflict that tip the scales despite the added risks of exercising that curiosity, but not necessarily for all individuals all of the time.

Given the added survival risks of curiosity and the fact that a relative few can provide shared benefits across the group, evolution has arguably seen fit to make a sizeable percentage of the population less adventurous, less curious, and somewhat conformist. This reduces their mortality rates by avoiding exploration and experimentation risks while also ensuring that previously established knowledge and techniques are well leveraged and handed down from generation to generation. These represent the benefits of conforming and, arguably, form part of the reason we see such variety from person to person. As with ant colonies, human colonies also hatch individuals who are somewhat predisposed to certain specializations, albeit not to the extreme extent seen in ants.

The highly curious tend to be relatively nonconformist, preferring to try their own techniques, choosing change over comfort. As George Bernard Shaw wrote, "The reasonable man adapts himself to the world; the unreasonable one persists in trying to adapt the world to himself. Therefore all progress depends on the unreasonable man."[57]

Change requires divergence, as alluded to in the 2014 movie, *Divergent*, though not necessarily in an antisocial or antiestablishmentarian way. With divergence, some risk must be borne. Nature may have reacted to this conundrum by creating variety, safeguarding the many while putting others in harm's way for the greater good. Many curious individuals risk their lives in their endeavors as evidenced by the executions of d'Ascoli, Servetus, Bruno, and de Dominis, with some torture thrown into the mix.

In 1626, Francis Bacon died of pneumonia while conducting scientific experiments. The great English philosopher, statesman, writer, and scientist was Lord Chancellor of England and hailed as the father of the formalized scientific method. He wrote poetry and plays, and to this day there remains controversy that he may have anonymously co-authored many of Shakespeare's plays, including Hamlet. An integral figure of the early Enlightenment and the preceding Scientific Revolution, he contracted pneumonia while, of all things, stuffing snow into a chicken as an experiment in refrigeration. Any activity out of the norm, even one seemingly innocuous and humdrum as this, bears risk simply by virtue of being out of the norm.[58]

While some died in pursuit of their passions, others prevailed to push the tribe and humanity to new heights. These are the Bantings, Jenners, and Pasteurs of the world. Not all members of the tribe need to take part in a discovery for the entire group to benefit. As a result, genetic variety may have favored a particular percentage of nonconformists or, as Shaw described them, unreasonable individuals. This percentage could perhaps better maximize survival chances overall and on average. For example, if the mortality rate in being highly curious and adventurous is two or three times greater for the adventurous individual and yet the mortality rates overall for members of the tribe may be reduced by fifty percent through discoveries and advancements made by them, it could turn out that a 70-30 percentage split of conformists to nonconformists would be optimal to maximize average survival rates across all members. This mix would then represent the ESS or Evolutionarily Stable Strategy that natural selection would favor.[59]

Unfortunately, given the number of variables involved, such as the benefits of other specialized roles within the community, it is difficult to realistically predict optimal percentages using theory alone. Nonetheless, such variety likely exists. Perhaps we have informally seen it in our daily lives. As Aristotle wrote in Poetics IV, "Understanding [manthanein] gives great pleasure not only to philosophers but likewise to others too, though the latter have a smaller share [emotional stake] in it."

As well as shared benefits, another reason that the percentage of curious individuals need not be extremely high is the law of diminishing returns. Any additional curious members may provide a diminishing incremental benefit to the group. Adding more cooks may spoil the broth, so to speak. Curiosity may then represent an unnecessary risk at times. A drop-off in overall survival chances may occur as the percentage of highly curious individuals exceeds a certain amount. They may step on each other's toes, duplicate effort unnecessarily, or enter analysis paralysis.

THE UNREASONABLE MINORITY

The nature of each potential advancement also plays a role. Some changes require a natural amount of time to "bake in." Throwing more people at a problem may not produce faster results when natural bottlenecks necessitate the passage of time before one discovery can morph into the next, incrementally. This effect has been seen in the software industry and elsewhere where adding to research and development staff may fail to improve progress due to natural inertia. Progress often requires time-consuming phases of testing, integration, assimilation, and experimentation. As Elon Musk said, "It is a mistake to hire huge numbers of people to get a complicated job done. Numbers will never compensate for talent in getting the right answer (two people who don't know something are no better than one), will tend to slow down progress, and will make the task incredibly expensive." This effect may also exist in nature as in the corporate world.

Another fundamental source of diminishing returns comes from the need for specialized roles beyond just degrees of curiosity. In an ant colony, there are workers, soldiers, and those dedicated to rearing young. Without a particular mix of such specialists, survival would not be optimally assured. The mix represents the ESS of the ant. While primates and humans do not specialize to this extent, some role-centricity clearly exists. We have had hunters and gatherers in the past. We have clearly had soldiers and caretakers. Distracting all members of the tribe with curiosity could be detrimental to such roles and, therefore, the overall success of the tribe. In the corporate world, to continue the analogy, it would be detrimental for the entire firm to be in the research department when marketing, manufacturing, and distribution are required.

Due to modern society and paid salaries, it is reasonable to suggest that curiosity would be universally beneficial today, driving academic and financial success. As a species, we are nonetheless stuck with whatever genetic variety natural selection has seen fit to apportion out based on millions of years of prehistory. Despite the demand for technologists, as a society we have a deficiency in the number of individuals choosing careers in technology and engineering. We still have free will, of course, to forge career plans, but variety manifests itself in our desires, aspirations, and skills. The ESS of the past, having been put in place over the course of millennia, continues to play out today. The same variety and hence specialization that provided benefits in tribal times may not be optimal today, but our genetic variety will only change slowly over multiple generations, if ever. Variance remains part of our species and will persist for generations before natural selection might orchestrate a beneficial change, such as creating a greater percentage of prospective engineers.

Any potential future change in our natural-born variety is hard to predict, further complicated by the fact that natural selection is no longer *natural*. Our very civilization is impacting our genetic future through family planning and other forms of highly unnatural selection. What took 2.3 million years to put in place is not easily eradicated. At the same time, the rules of genetic selection have now changed, taking on new dynamics never before seen on earth.

Regardless of the optimal percentage of curious individuals, the rule of shared benefits remains. The few bring benefits to the many, but not without resistance. Once a curious individual makes a discovery, the advancement may not immediately be met with open arms. People are generally resistant to change, especially when the group includes conformists who, by nature, lean towards the status quo. This is not necessarily a bad thing. A little resistance to change is healthy. It may serve to validate the advancement and avoid potentially frivolous or detrimental changes.

A fringe erratic, as a curious individual can often appear, is not necessarily to be taken at his or her word, whether human or primate. An untested and unproven change could turn out to be problematic. In prehistoric times, it could lead to the slaughter of the group. Nature may have chosen to address this problem by creating a skeptical phase before a change is adopted. This effect was painfully demonstrated by the British Royal Society in rejecting Jenner's smallpox vaccine and by Felix Pouchet in refuting Pasteur's theories on bacteria.

An ironic example of conformism and resistance to change can be seen in the United States public educational system, bordering on negligence. While technology is omnipresent in society and within the workforce, training in software and technology architecture, today, is not mandatory in middle or high schools. As a result, college major selection by individuals is based on imperfect information. Due to a lack of experience in basic technologies and since individuals necessarily choose topics based on knowledge possessed at that time, which is greatly influenced by societal norms and a dated educational system, this leads to a shortage of engineers and a struggling local workforce with large student loans to repay for those unfortunate enough to choose a specialty for which demand is lacking.

Regardless of such challenges, despite resistance to change and the fact we need not all be intensely curious, changes for the greater good do eventually occur. Advancements are ultimately shared across the group for the betterment of all. Progress is made due to the efforts of the salient few, the most curious among us, investing time and effort, occasionally risking life and limb but not to be martyrs. They are simply driven.

These are the pretraumatized among us, pre-programmed with pleasures and pains that impel them to be divergent. They personify an evolutionary force that made us who we are today, and what we are: human. To paraphrase Shakespeare's *Henry V*, these are the few, the happy few, the band of brothers and sisters, for those today who shed their blood with me will be my kin, we, the unreasonable minority.

4.3 A Happiness Dilemma

The road to happiness is a personal one for which the direction is not always clear at the best of times. When curiosity is thrown into the mix, things can appear extremely paradoxical. For the highly curious, the road to happiness includes hairpin turns, is undulating, and is intensely awkward to navigate. Any other road, such as a smooth, straight, and predictable road with a nice view of rose gardens and friends waving, while ideal for some, would render such people bored, frustrated, or restless over time.

HEAVEN CAN BE AN EVENTUAL HELL.

Struggle combined with an occasional triumph defines the highly curious and their form of happiness, such as it may be. For them, happiness often involves prolonged periods of pain amid hard work followed by satisfied reflection when an incremental benefit is achieved. They experience a momentary appreciation of how worthwhile it all was in the end. With hindsight, they may recall how alive they felt in their struggle while it may have simply felt like a sacrifice in the moment. There would be the loss of quality time to spend with friends and loved ones, the loss of freedom to watch meaningless TV shows, and a voluntary exile from the comfortable mundane. Curiosity is a self-imposed sentence passed by a deep, forgotten part of the psyche.

The movie *Whiplash* depicts an analogous effect wherein an abusive teacher and an abused student find themselves, in fact, symbiotically connected in their own personal struggles. They each hope these struggles will culminate in a memorable, unique moment. They each need the other, each equally frustrated but garnered by a belief that the end justifies the means, that upon reflection the endured pain will indeed be reinterpreted as joy, the achievement all the merrier for the torment that came before. This effect is an echo of the temporality of curiosity itself, justifying its struggle only over time, in retrospect, arriving at some unknown, pivotal future with something akin to joy and ecstasy, a form of love beyond love.[60]

Does one moment of achievement make up for a day, a month, or a year of pain and struggle? Perhaps this is the wrong question to ask if we acknowledge that the pain and struggle are part of the glory. Success is not merely a compensation for the preceding pain. The success renders the pain beautiful as if the achievement makes us reinterpret the pain in a positive light, in reflection. There is nothing good or bad but thinking that makes it so, to paraphrase Shakespeare, for the hard work and effort, rather than things to look back on in anguish, become glory epitomized, part of being engaged in life itself. It is about leaving a town called Status Quo on a long and arduous journey to Who Knows Where.

With this dichotomy in mind, meaning the relationship between current pain and later satisfaction, the road to personal happiness can be an unexpected one. As well as leading to intelligence, achievement, and success, curiosity may also pave the way to an unexpected form of happiness involving toil that, in turn, requires perhaps a degree of courage to pursue. In a modern world, filled with distractions that our ancestors never had the detriment to deal with and that we did not have an opportunity to evolve intuitive mechanisms for, extreme self-awareness is required in order to seek out and launch fulfilling endeavors. Without such self-awareness, the media, marketing, games, shows, gameshows, and other forms of entertainment are well positioned to make decisions on our behalf. Such modern elements can impose themselves unrewardingly on the curious.

Knowing the origin of curiosity may lead us to believe we can ignore it and thereby dispense with it, now that we are not in such great peril. The need to exercise our curiosity has diminished, given that death rates from preventable causes at the hands of enemies or predators are not what they used to be. We are nonetheless victims of our own success, victims emotionally, that is. We can only ignore our emotional programming at our peril, putting ourselves at risk of being ill at ease on a recurring basis, flirting with the danger of an enduring discontent.

We are living in a world where levels of depression and anxiety are far higher than a hundred years ago despite life being far more comfortable than ever. According to the World Health Organization, 350 million people suffer from depression in what it describes as one of the biggest epidemics the world has known. In a 2015 study of 100,000 students attending colleges across the United States, the Center for Collegiate Mental Health found that one in every three students had taken psychiatric medication, one in four had seriously considered suicide, one in ten had been hospitalized for psychiatric reasons, and one in ten had actually attempted suicide. It cited anxiety and depression as the primary causes as assessed by college counselors.[61]

According to the National Institute of Mental Health, one in three girls and one in five boys between the ages of thirteen and eighteen suffer from generalized anxiety disorder with many receiving prescriptions of Xanax or equivalent *anxiolytics*. Such drugs are essentially tranquilizers with side effects that include memory problems, confusion, and, ironically, depressed moods and suicidal thoughts. Given greater personal freedom and opulence than at any other period of human history, clearly something is amiss.[62]

We are surrounded by television, websites, and apps that eagerly accept our attention, and yet these represent backup or alternate activities. We do not consciously choose them. They choose us, in a sense. Technologists refer to backup or alternate modes as "defaults." Virtually all complex software has default modes of operation, or default configuration settings. The curiosity app is no exception.

Devoid of a consciously chosen goal or activity, curiosity will react to whatever happens to be in the immediate vicinity, inciting mental activity to analyze the minutiae of trivialities through haphazard streams of consciousness. In nature, this was a reasonable default mode of operation given our ancestors' exposure to ecological surroundings that must always be dealt with. In modernity, it can invite emotional catastrophe. We did not evolve around our newly fabricated civilization, hence we lack intuitive mental apparatus for peace of mind in its confines.

In modernity, marketing, word of mouth, and the media may subtly select activities for us—mental white noise to no rewarding end—if we do not invest preliminary contemplation up front to gain conviction in alternative tasks. News of social injustices or environmental issues can readily take an exaggerated footing in the adolescent psyche, for instance, contributing to teen angst. Embarrassing tiffs on social media are potentially elevated to soul-crushing status in our brave new technological world. On the other hand, seeking out and pursuing rewarding endeavors requires conscientious attention. Curiosity now needs to be channeled like never before if discontent is to be avoided and more enduring emotional benefits are to abound.

In studies performed on depression and suicide, lack of motivation is often cited as a dominant factor rather than a specific sadness or pain. While curiosity is an intrinsic motivation, a course nevertheless needs to be plotted. It does not always select the best route, particularly given the abundance of modern-day distractions. The driver app occasionally needs to be put into gear.

Research performed by James Rodrigue goes one step further, highlighting a specific link between moods and levels of curiosity. Individuals experiencing depressed or saddened states would consistently become less curious based on their measured desires to acquire information and perceptions of how useful or important that information was. This appeared in stark contrast to results seen in individuals experiencing neutral, joyous, or exhilarated states, where such individuals remained relatively curious. It may, therefore, be theorized that happiness feeds curiosity, which in turn feeds happiness in a positive reinforcement loop, albeit with hard work and vigorous effort thrown into the mix. This positive reinforcement gives cause for optimism, but only with effort and conscious, upfront deliberation.[63]

The relationship between curiosity and happiness provides a clue of curiosity's hidden past. Happiness for the curious is intertwined with struggle and the ability to self-motivate toward that struggle. For some, achieving contentment involves an imperative to strive and advance. This is a profound manifestation of curiosity's primal roots. We are armed prosthetically and cranially with a tool for advancement, a mental mechanism to invoke work, an insatiable need. It not only triggered our evolution into this particular mammalian form, Homo sapiens *sapiens*, it is also integral to personal happiness.

While the apparent relationship between curiosity and happiness is not direct proof that curiosity has been a defining force of human natural selection, it is powerful circumstantial evidence and continues to form a pattern of inference. The importance of curiosity-oriented activity in daily life, paired with hard work amid occasional accomplishments, illustrates a profound link between our evolved feelings and their purpose. They speak of our innumerable past evolutionary interactions that benefitted from such upfront activity. To rest on one's laurels was to await destruction in primordial times. Today, to ignore our underlying curiosity is to welcome malaise.

With the appropriate degree of homage paid to the curiosity app by plotting a course, individuals may find greater fulfillment. This is particularly feasible when the iterative aspect of curiosity-oriented endeavors is considered. Hard work may produce only minor results, initially. When learning a new skill, starting a new field of study, or launching a hobby, incremental gains may be modest in isolation, but eventually compound. These seemingly innocuous improvements can add up, one building upon the other. Correspondingly, the software industry includes a phenomenon known as iterative development. Early versions of software seek to satisfy design goals but only in a simplistic manner, initially. Later versions incrementally build upon the simpler early versions. The cycle continues from one software generation to the next. Analogies exist in curiosity-oriented pursuits whereby each milestone or accomplishment, while potentially insignificant in isolation, can eventually lead to greater things, spurring us on iteratively—but unrelentingly.

It should perhaps come as a relief to not have to feel happy all the time, to not feel such stereotypical social pressure. The United States' constitutional right to pursue happiness is just that, a pursuit. It is a journey, not a destination, in a very real sense. Humans, particularly those most curious, are not necessarily destined to be happy and joyous on a continual basis. Such a state would ultimately feel confining for such individuals. Instead, struggle, work, challenge, focused determination towards goals, and incremental achievement amid occasional respites are the names of the game. Devoid of pressure to have to always feel happy or else worry that something is amiss or wrong with us, the highly curious may thus find their deepest form of solace. Perhaps herein lies one source of happiness, after all, namely the lack of expecting it.

Milestoning potentially aids this process, rewarding us emotionally along the way, whether at the end of a productive day or following a breakthrough. Feelings of discomfort dissipate, and satisfaction is momentarily felt. In classical psychiatry, these feelings can be somewhat confusing since they come and go, oscillating as more effort is soon invested, a thirst that refuses to be quenched. Indeed, such is the curiosity drive, geared towards the long-term good. The more intrinsic the curiosity, the further out and more uncertain the goal. Satisfaction is, therefore, elusive and happiness occasionally confusing. Curiosity is a teasing drive, somewhat rewarding but also somewhat dissatisfying, stringing us along insatiably, amid effort and work, amid focused engagement, one step to the next. Curious individuals are in a constant state of running towards happiness and never quite getting there—a precarious state. Perhaps that is the manner in which curious individuals are most happy: engaged insatiably in a curiosity paradox.

While it may require hard work to consciously feed and nurture one's curiosity—to consistently channel its raw energy—the *pursuits* of curiosity are often inextricably linked to our wellbeing. They can arguably lead to happiness and fulfillment. Curiosity shaped our evolution for over two million years and, analogously, shapes the evolution of countless technologies and hence quite lucrative technology jobs. Given advances in medicine that are often the result of curiosity, it can help us live long and prosper, as Star Trek's Spock might say, but work is involved. Spurred on in no small part by frustration, dissatisfaction, and boredom-avoidance, the pursuit of happiness through curiosity-oriented endeavors requires effort and discipline.

Ultimately, there are rewards. Looking back it is possible to revel in one's achievements and, perhaps more profoundly,

A HAPPINESS DILEMMA

the recollection of the journey. As painful or strenuous as it may have seemed at the time, in retrospect, one recalls the vitality of the whole process, the feeling of being engaged in life, of not letting time or opportunities pass one by, and of being alive. It is the reward for pursuing one's true nature and perhaps our most virtuous of traits.

When urgency appears in the highly curious, it may occasionally seem irrational. Some individuals may seem socially inept or rub people the wrong way. In extreme cases, friendships may be lost. Such is the frustration and dissatisfaction that the highly curious occasionally personify. Sanity and sociability may indeed be relative, as Ray Bradbury alluded to, but energy is not. It is absolute. It either exists or does not. While there may be differences in the degree of political correctness or geniality from one individual's approach to the next, there is no doubting the hallmark of animation or drive. Energy is supreme—and intrinsic motivation its source.

Energy imparts activity where there otherwise might be none, other than resting and relaxing as might a well-fed mammal in the woods. Agreeability and niceness are not necessarily signatures of curiosity, as expected given its violent birth. Curiosity may occasionally manifest toil and frustration. Individuals may strive for activity even when others might get in the way, pushing through the crowds of naysayers, launching headlong with no thought of compromise, no intention of taking no for an answer. After all, lives once depended on it.

In the comfort of modernity, we may not obviously be in danger. We nonetheless remain curious. We carry the baton of curiosity in the evolutionary relay race passed from

Australopithecus to Homo habilis to Homo erectus to us. The urgency of curiosity cannot be seen, but it is understood subconsciously, deep down in a forgotten part of the brain. It is the ultimate survival tool. It ensured that neurons would be effectively used that might otherwise be a burden, allowing the brain to evolve. Curiosity was once a matter of life and death. For some, even today, it maintains that familiar signature, emotionally. Without being true to our curiosity, we cannot truly feel alive.

Notes

1. D. Goleman, *Emotional Intelligence: Why It Can Matter More Than IQ* (Bantam Books, September 27, 2005). National Institute of Mental Health, Merikangas, He, Burstein, Swanson, Avenevoli, Cui, Benjet, Georgiades, Swendsen, *Lifetime prevalence of mental disorders in U.S. Adolescents* (July 31, 2010). Penn State, Center for Collegiate Mental Health, *2014 Annual Report, 2015 Annual Report*. Penn State News, *Annual report offers snapshot of U.S. college students' mental health needs* (February 5, 2015). World Health Organization, World Federation for Mental Health, *DEPRESSION: A Global Crisis* (October 10, 2012)
2. Charles Darwin, *On the Origin of Species by Means of Natural Selection, or the Preservation of Favoured Races in the Struggle for Life* (London, John Murray, 1859).
3. D. D. Clark & L. Sokoloff, *Basic Neurochemistry: Molecular, Cellular and Medical Aspects*, eds. Siegel, G. J., Agranoff, B. W., Albers, R. W., Fisher, S. K. & Uhler, M. D. (Lippincott, Philadelphia, 1999), pp. 637-670.
4. Central Intelligence Agency, *The World Factbook*, Country Comparison :: Infant Mortality Rate: https://www.cia.gov/library/publications/the-world-factbook/rankorder/2091rank.html; Central Intelligence Agency, *The World Factbook*, Country Comparison :: Maternal Mortality Rate: https://

www.cia.gov/library/publications/the-world-factbook/rankorder/2223rank.html.

5. B.L. Beyerstein, *Mind Myths. Exploring Popular Assumptions about the Mind and Brain: Whence Cometh the Myth that We Only Use 10% of Our Brains?* (Chichester: John Wiley and Sons, 1999).
6. See mirror.
7. The Guardian, Robin McKie, *Colin Blakemore: how the human brain got bigger by accident and not through evolution* (March 27, 2010). Ian Tattersall, *An evolutionary framework for the acquisition of symbolic cognition by Homo sapiens* (2008). Comp Cogn Behav Revs 3: 99-114. Illustration Gisselle Garcia.
8. Scientific American, Rachael Moeller Gorman, *Cooking Up Bigger Brains* (December 16, 2007).
9. Ralph Waldo Emerson, *Essays: First Series*, *Self-Reliance* (1841).
10. Charles Darwin, *The Descent Of Man, And Selection In Relation To Sex* (London, John Murray, 1871).
11. The Guardian, Neuroscience, Mo Costandi, *Bipedalism, birth and brain evolution* (May 7, 2012).
12. Robin Dunbar, *Neocortex Size As A Constraint On Group Size In Primates, Journal of Human Evolution (1992).* Journal of Human Evolution, doi:10.1016/0047-2484(92)90081-J.
13. Terence McKenna, *Food for the Gods* (Bantam, 1993); Julian Jaynes, *The Origin of Consciousness in the Breakdown of the Bicameral Mind* (Houghton Mifflin, 1990).
14. Robert A. Rohde, Global Warming Art Project, Wikimedia Commons, *Five Million Years of Climate*

Change from Sediment Cores (December 31, 2004); based on data from Lisiecki and Raymo.

15. Time, Barb Darrow, *Bezos Says Amazon Web Services Is a $5 Billion Business* (April 23, 2015).
16. Charles Pasternak, *What Makes Us Human: Curiosity and Quest* (Oneworld Publications, September 2007). G. Roth, U. Dicke, *Evolution of the brain and intelligence* (May 2005). Trends Cogn. Sci. (Regul. Ed.). 9 (5): 250–7. PMID 15866152. doi:10.1016/j.tics.2005.03.005.
17. Alison Gopnik, Andrew N. Meltzoff & Patricia K. Kuhl, *The Scientist in the Crib: What Early Learning Tells Us About the Mind* (William Morrow Paperbacks, December 26, 2000); TED talks, www.ted.com, Alison Gopnik, *What Do Babies Think?* (October 2011).
18. Cristine H. Legare & Jennifer M. Clegg, *The Development of Children's Causal Explanations* (The University of Texas at Austin, 2015).
19. Museum of Science and Industry, Manchester, England, *The Baby: the World's First Stored-Program Computer* (www.mosi.org.uk).
20. George Loewenstein, *The Psychology of Curiosity: A Review and Reinterpretation* (The American Psychological Association, 1994, http://dx.doi.org/10.1037/0033-2909.116.1.75).
21. Gregory Berns, *Satisfaction: The Science of Finding True Fulfillment* (Henry Holt and Co., 2005).
22. Max-Planck-Gesellschaft, Munich, Germany, *"Personality-Gene" makes Songbirds Curious* (May 2, 2007).

23. TED Conferences, www.ted.com, David Deutsch, *A new way to explain explanation* (October, 2009).
24. University of Michigan, Electrical Engineering and Computer Science; Singh, Barto & Chentanez, *Intrinsically Motivated Reinforcement Learning* (2010).
25. Walter Mischel, *Cognitive and attentional mechanisms in delay of gratification* (Journal of Personality and Social Psychology, 1972, PMID: 5010404).
26. The New York Times, Science, John Noble Wilford, *Fossils in Kenya Challenge Linear Evolution* (August 9, 2007).
27. Science Magazine, Bert Holldobler, *Tournaments and Slavery in a Desert Ant* (28 May, 1976).
28. Journal of Mammalogy, L. David Mech, *Buffer Zones Of Territories Of Gray Wolves As Regions Of Intraspecific Strife* (February 1994).
29. Jane Goodall, *The Chimpanzees Of Gombe: Patterns Of Behavior* (Harvard University Press, 1986). Richard Wrangham and Dale Peterson, *Demonic Males: Apes and the Origins of Human Violence* (Mariner Books, November 14, 1997).
30. Yearbook of Physical Anthropology, Richard W. Wrangham, *Evolution of Coalitionary Killing* (1999). Harvard University, www.fas.harvard.edu, Richard Wrangham, Michael Wilson & Martin Muller, *Comparative Rates Of Violence In Chimpanzees And Humans* (20 July, 2004).
31. Edinburgh University, Konrad Lohse and Wageningen University, Laurent Frantz: News Wise, Genetics Society of America, *New Method Confirms Humans and Neanderthals Interbred* (April 8, 2014).

32. Lawrence H. Keeley, *War Before Civilization: The Myth of the Peaceful Savage* (Oxford University Press, December 18, 1997).
33. Smithsonian, Larry Zimmerman & Richard Whitten, *Mass Grave at Crow Creek in South Dakota Reveals How Indians Massacred Indians in 14th Century Attack* (September 1980).
34. Richard Wrangham and Dale Peterson, *Demonic Males: Apes and the Origins of Human Violence* (Mariner Books, November 14, 1997).
35. Thomas Hobbes, *Leviathan, or The Matter Forme & Power of a Commonwealth, Ecclesiasticall and Civill* (Cambridge: at the University Press, 1904).
36. Raymond Dart, American Journal of Physical Anthropology, *The predatory implement technique of the Australopithecines* (1949). Robert Broom, John Robinson & Raymond Dart, Swartkrans Cave (Gauteng, South Africa), located ~40 km northwest of Johannesburg, in Gauteng Province, South Africa, www.swartkrans.org.
37. Arthur C. Clarke, *2001: A Space Odyssey* (Hutchinson, 1968).
38. Robert Ardrey, *African Genesis* (Atheneum, 1961).
39. J. D. Pruetz, P. Bertolani, *Savanna chimpanzees (Pan troglodytes verus) hunt with tools* (2007, doi:1016/j.cub.2006.12.042). J. D. Pruetz, P. Bertolani, K. Boyer Ontl, S. Lindshield, M. Shelley, E. G. Wessling, *New evidence on the tool-assisted hunting exhibited by chimpanzees (Pan troglodytes verus) in a savannah habitat at Fongoli, Senegal* (April 15, 2015, DOI: 10.1098/rsos.140507).

40. Journal of Taphonomy, Geoff M. Smith, *Damage Inflicted Upon Animal Bone by Wooden Projectiles: Experimental Results Andarchaeological Implications* (Prometheus Press/ Palaeontological Network Foundation, 2003).
41. Hillard Kaplan, Kim Hill, Jane Lancaster, A. Magdalena Hurtado, *A Theory of Human Life History Evolution: Diet, Intelligence and Longevity* (2000); Evolutionary Anthropology; 9 (4): 156–185. doi:10.1002/1520-6505(2000)9:4<156::AID-EVAN5>3.0.CO;2-7.
42. Y. Fernandez-Jalvo, J. C. Díez, I. Cáceres, & J. Rosell,*Human cannibalism in the Early Pleistocene of Europe (Gran Dolina, Sierra de Atapuerca, Burgos, Spain)* (September 1999). Journal of Human Evolution (Academic Press) 37 (34): 591-622. doi:10.1006/jhev.1999.0324.
43. R. Zhu, et al, *Earliest presence of humans in northeast Asia* (2001). Nature 413, 413-417.; R. Zhu, et al, *New evidence regarding the earliest human presence at high northern latitudes in northeast Asia* (2004). Nature 431, 559-562.; R. Zhu, et al, *Early evidence of the genus Homo in East Asia*, Journal of Human Evolution (2008). 55, 1075-1085. D. Lordkipanidze, et al, *A Complete Skull from Dmanisi, Georgia, and the Evolutionary Biology of Early Homo*, Science (October 18, 2013): Vol. 342, Issue 6156, pp. 326-331: DOI: 10.1126/science.1238484. Science Magazine, *Meet the frail, small-brained people who first trekked out of Africa* (2016): http://www.sciencemag.org/news/2016/11/

meet-frail-small-brained-people-who-first-trekked-out-africa.
44. G. Holstege, et al: *Brain activation during human male ejaculation* (2003), The Journal of Neuroscience. B. Komisaruk, B. Whipple: *Functional MRI of the Brain During Orgasm In Women* (2005). Thomas R. Kosten, Tony P. George: *The Neurobiology of Opioid Dependence: Implications for Treatment* (2002), Sci Pract Perspect, National Center for Biotechnology Information, U.S. National Library of Medicine.
45. "Insanity is relative. It depends on who has who locked in what cage." Raymond Bradbury, *The Meadow* (1947), originally a radio play for the World Security Workshop; later revised as a short story.
46. World Health Organization, *Factsheet*, Media center: http://www.who.int/mediacentre/factsheets/fs310/en/
47. Richard Wrangham and Dale Peterson, *Demonic Males: Apes and the Origins of Human Violence* (Mariner Books, November 14, 1997).
48. Attribution of this quote to Mark Twain cannot be verified.
49. Thomas Aquinas, *Summa Theologica* (1471).
50. Northwestern University, Neuroscience and Robotics Lab, https://nxr.northwestern.edu/people/malcolm-maciver ; PNAS, Malcolm A. MacIver, Lars Schmitz, Ugurcan Mugan, Todd D. Murphey, Curtis D. Mobley: *Massive increase in visual range preceded the origin of terrestrial vertebrates*, vol. 114 no. 12 > E2375-E2384, doi: 10.1073/pnas.1615563114.

51. Yearbook of Physical Anthropology, Richard W. Wrangham, *Evolution of Coalitionary Killing* (1999).
52. Eds F. Coward, R. Hosfield, M. Pope, F. Wenban-Smith, Settlement, *Society and Cognition in Human Evolution: Landscapes in the Mind* (Cambridge University Press, 2015).
53. National Institutes of Health, US National Library of Medicine, Stefan Riedel, *Edward Jenner And The History Of Smallpox And Vaccination* (January 2005).
54. Gerald L. Geison, *The Private Science of Louis Pasteur (Princeton Legacy Library)* (Princeton University Press, April 17, 1995). The College of Physicians of Philadelphia, historyofvaccines.org.
55. Michael Bliss, *Banting: A Biography* (University of Toronto Press, March 15, 1993). *The Young Doctor Intimately Knew Despair*, Anal. Chem., 1955, 27 (9), pp 19A-19A, DOI: 10.1021/ac60105a721, Publication Date: September 1955.
56. Robin Marantz Henig, *The Monk in the Garden: The Lost and Found Genius of Gregor Mendel, the Father of Genetics* (Mariner Books, May 12, 2001).
57. George Bernard Shaw, *Man and Superman* (1903).
58. Michael Peppiatt, *Francis Bacon: Anatomy of an Enigma* (Skyhorse Publishing, September 1, 2009)
59. Maynard Smith, J., Price, G.R., 1973. *The logic of animal conflict*. Nature 246, 15-18 (doi:10.1038/246015a0).
60. *Whiplash* (2014), written and directed by Damien Chazelle (Bold Films, Blumhouse Productions, Right of Way Films).

NOTES

61. World Health Organization, World Federation for Mental Health, *DEPRESSION: A Global Crisis* (October 10, 2012). Penn State, Center for Collegiate Mental Health, *2014 Annual Report* and *2015 Annual Report*. Penn State News, *Annual report offers snapshot of U.S. college students' mental health needs* (February 5, 2015). BBC News, *Depression looms as global crisis* (September 2, 2009).
62. National Institute of Mental Health, Merikangas, He, Burstein, Swanson, Avenevoli, Cui, Benjet, Georgiades, Swendsen, *Lifetime prevalence of mental disorders in U.S. Adolescents* (July 31, 2010).
63. J. Rodrigue, K. Olson, R. Markley, *Induced mood and curiosity* (1987), Cogn Ther Res 11: 101. doi:10.1007/BF01183135

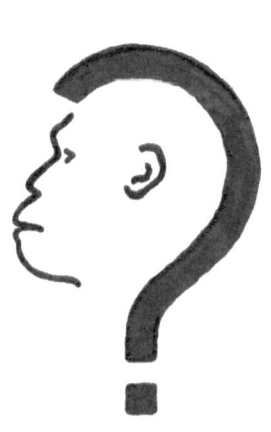

Index

2001: A Space Odyssey, 60
Afghanistan, 6, 109
algorithm, xi, 33, 75, 114
Amazon, 23, 105
anthrax, 130, 131
ants, 51, 112, 113
anxiety, 150
anxiolytics, 150
Aquinas, Thomas, 108
Ardipithecus, 48, 54
Ardrey, Robert, 60, 61
Aristotle, 131, 142
Australopithecus genus, 3, 43, 50, 68, 98, 139, 156
    Australopithecus garhi, 48
Bacon, Francis, 141
Banting, Frederick, 132, 133, 134
Bezos, Jeff, 23
Blakemore, Colin, 10
bonobo, 44
Bordain, Anthony, 12
Bradbury, Ray, 85, 155
breadth curiosity, 35
Brin, Sergey, xi
British Medical Journal, 132
cannibalism, 51, 66

Captain Cook, 128
Carnegie, Dale, 8
catch 22. *See* Hobbesian trap
Center for Collegiate Mental Health, 149
Central Intelligence Agency, 6
cholera, 130, 131
choppers and scrapers, 50, 63
Clarke, Arthur C., 60
cognition theory, 37
cognitive psychology, 39
computer science
    defaults, 150
    driver, 23, 103, 113
    iterative development, 153
Copernicus, Nicolaus, 14, 135, 136, 137
curiosity, defined, 32
Dart, Raymond, 60, 61
Darwin, Charles, 2, 4, 14, 16
d'Ascoli, Cecco, 141
delayed gratification, 40
Demonic Males, 93
depression, 149
depth curiosity, 35
Deutsch, David, 37, 40, 41

diabetes, 132
Divergent movie, 10
DRD4 gene, 35
drive theory, 34
Dunbar, Robin, 16
Earth Liberation Front (ELF), 88
Einstein, Albert, x, 137
Emerson, Ralph Waldo, 13, 14
emotion, 81
encephalization, 4, 16, 46, 114
Enlightenment, 38, 136, 141
environmental conservation, 88
Ethiopia, 48
evolutionary psychology, 39, 43, 90
evolutionary side effect, 90
Expensive Tissue Hypothesis, 12, 13
Falk, Dean, 15
Fleming, Alexander, 104
Galilei, Galileo, 14, 135
game theory, 58
genocide, 52, 53, 56
Global Positioning System (GPS), 104
Gombe, 52, 53, 118
Goodall, Jane, 52

Google, xi, 103
Gopnik, Alison, 26, 68, 70, 71, 87
Gran Dolina, Spain, 66
Haffkine, Waldemar, 131
heart disease, 92
heroin, 82
Homo genus, 18, 49, 60
    Homo antecessor, 66
    Homo erectus, 49, 156
    Homo habilis, 18, 43, 50, 156
    Homo heidelbergensis, 55
    Homo sapiens, 3, 55, 121
    Homo sapiens neanderthalensis, 56
    Homo sapiens sapiens, 56
incongruity theory, 35
infanticide, 51, 52
inoculation, 129
Inquisition, 135, 136
internet, 104
intrinsic curiosity, defined, 38
Jaynes, Julian, 18
Jefferson, Thomas, 129
Jenner, Edward, 127, 128, 129, 130, 134, 145
Kahama, 53
Keeley, Lawrence, 56
kidnapping, 50, 51, 53
Kilburn, Tom, 34

INDEX

Kubrick, Stanley, 60
Legare, Cristine, 26, 68, 70, 87
Lucy movie, 8
MacIver, Malcolm, 114
macroevolution, 10
magic mushrooms. *See* psilocybin
Manchester, x, 34
marshmallow test, 40
McKenna, Terence, 18
Mech, Lucyan, 52
Mendel, Gregor, 134, 136
microwaves, 104
Mischel, Walter, 40
Musk, Elon, 100, 101, 143
narcotics, 82
Nash equilibrium, 58
National Institute of Mental Health, ix, 150
Neanderthal, 55, 56, 121
New Guinea, 57
Newton, Isaac, 14
Occam's razor, x
Ockham, William of, x
opiods, 82
Orrorin, 54
overeating, 83
ovicide, 52
Page, Larry, xi
Paleolithic, 64, 66
palisade, 56
Pasteur, Louis, 130, 131, 132, 134, 145
pasteurization, 131
penicillin, 104
Peterson, Dale, 65, 93
Peterson, Richard, 57
Posttraumatic Stress Disorder (PTSD), 84, 85, 86, 87, 89
Pouchet, Felix, 131, 145
Power, Margaret, 118
prepping, 87, 88
pretraumatic stress, 86
psilocybin, 18
public education, 145
rabies, 130, 131
rape, 53, 56
Rodrigue, James, 151
Roebroeks, Wil, 121
Schiller trap. *See* Hobbesian trap
scientific method, 69, 141
Scientific Revolution, 141
Servetus, Michael, 135, 141
Shakespeare, William, 123, 141, 146, 148
Shaw, George Bernard, 141, 142
slavery, 50, 51, 56
smallpox, 127, 128, 129, 130, 145

Smith, Geoff, 64
Social Brain Hypothesis, 16
Somalia, 6
spandrel, 90
spontaneous generation, 131
sport, 89, 90, 92
Stockholm syndrome, 51, 53
Stone Age, 64
Sudan, 6
suicide, 149
survivalism. *See* prepping
Tanzania, 52, 53, 118
trait curiosity, 35
tree resin, 41
tribes
   Mae Enga, 57
   Native American, 56
Yanomamo, 57
Turing, Alan, 34, 81
Twain, Mark, 100
vaccination, 127, 128, 130
Whiplash movie, 148
wolves, 52, 112
World Health Organization, ix, 92, 127, 149
world wars
   World War I, 71
   World War II, 53, 71, 80, 104, 127
Wrangham, Richard, 12, 57, 65, 93

Xanax, 150
Zollikofer, Christoph, 15

About the Author

Somewhere around the periphery of Lake Dunmore, VT

Alexandros S. Kourt is both a science writer and a computer engineer. He has over twenty years of experience designing and building large-scale computer systems and has been a contributor to *Psychology Today* since 2009. Of Greek Cypriot (and ape) descent, Kourt was born in Liverpool, England, and grew up in rural Yorkshire. He now divides his time between New York, Vermont, and Cyprus.

 www.curiositygene.com

Thank you for your interest in The Curiosity Gene. Comments and feedback can be sent directly to the author at info@curiositygene.com. Positive reviews on Amazon are most welcome and greatly appreciated!

www.ingramcontent.com/pod-product-compliance
Lightning Source LLC
Chambersburg PA
CBHW030939180526
45163CB00002B/632